Computable Calculus

LIMITED WARRANTY AND DISCLAIMER OF LIABILITY

ACADEMIC PRESS, INC. ("AP") AND ANYONE ELSE WHO HAS BEEN INVOLVED IN THE CREATION OR PRODUCTION OF THE ACCOMPANYING CODE ("THE PRODUCT") CANNOT AND DO NOT WARRANT THE PERFORMANCE OR RESULTS THAT MAY BE OBTAINED BY USING THE PRODUCT. THE PRODUCT IS SOLD "AS IS" WITHOUT WARRANTY OF MERCHANTABILITY OR FITNESS FOR ANY PARTICULAR PURPOSE. AP WARRANTS ONLY THAT THE MAGNETIC DISC(S) ON WHICH THE CODE IS RECORDED IS FREE FROM DEFECTS IN MATERIAL AND FAULTY WORKMANSHIP UNDER THE NORMAL USE AND SERVICE FOR A PERIOD OF NINETY (90) DAYS FROM THE DATE THE PRODUCT IS DELIVERED. THE PURCHASER'S SOLE AND EXCLUSIVE REMEDY IN THE VENT OF A DEFECT IS EXPRESSLY LIMITED TO EITHER REPLACEMENT OF THE DISC(S) OR REFUND OF THE PURCHASE PRICE, AT AP'S SOLE DISCRETION.

IN NO EVENT, WHETHER AS A RESULT OF BREACH OF CONTRACT, WARRANTY, OR TORT (INCLUDING NEGLIGENCE), WILL AP OR ANYONE WHO HAS BEEN INVOLVED IN THE CREATION OR PRODUCTION OF THE PRODUCT BE LIABLE TO PURCHASER FOR ANY DAMAGES, INCLUDING ANY LOST PROFITS, LOST SAVINGS OR OTHER INCIDENTAL OR CONSEQUENTIAL DAMAGES ARISING OUT OF THE USE OR INABILITY TO USE THE PRODUCT OR ANY MODIFICATIONS THEREOF, OR DUE TO THE CONTENTS OF THE CODE, EVEN IF AP HAS BEEN ADVISED ON THE POSSIBILITY OF SUCH DAMAGES, OR FOR ANY CLAIM BY ANY OTHER PARTY.

ANY REQUEST FOR REPLACEMENT OF A DEFECTIVE DISC MUST BE POSTAGE PREPAID AND MUST BE ACCOMPANIED BY THE ORIGINAL DEFECTIVE DISC, YOUR MAILING ADDRESS AND TELEPHONE NUMBER, AND PROOF OF DATE OF PURCHASE AND PURCHASE PRICE. SEND SUCH REQUESTS, STATING THE NATURE OF THE PROBLEM, TO ACADEMIC PRESS CUSTOMER SERVICE, 6277 SEA HARBOR DRIVE, ORLANDO, FL 32887, 1-800-321-5068. AP SHALL HAVE NO OBLIGATION TO REFUND THE PURCHASE PRICE OR TO REPLACE A DISC BASED ON CLAIMS OF DEFECTS IN THE NATURE OR OPERATION OF THE PRODUCT.

SOME STATES DO NOT ALLOW LIMITATION ON HOW LONG AN IMPLIED WARRANTY LASTS, NOR EXCLUSIONS OR LIMITATIONS OF INCIDENTAL OR CONSEQUENTIAL DAMAGE, SO THE ABOVE LIMITATIONS AND EXCLUSIONS MAY NOT APPLY TO YOU. THIS WARRANTY GIVES YOU SPECIFIC LEGAL RIGHTS, AND YOU MAY ALSO HAVE OTHER RIGHTS WHICH VARY FROM JURISDICTION TO JURISDICTION.

THE RE-EXPORT OF UNITED STATES ORIGINAL SOFTWARE IS SUBJECT TO THE UNITED STATES LAWS UNDER THE EXPORT ADMINISTRATION ACT OF 1969 AS AMENDED. ANY FURTHER SALE OF THE PRODUCT SHALL BE IN COMPLIANCE WITH THE UNITED STATES DEPARTMENT OF COMMERCE ADMINISTRATION REGULATIONS. COMPLIANCE WITH SUCH REGULATIONS IS YOUR RESPONSIBILITY AND NOT THE RESPONSIBILITY OF AP.

Computable Calculus

Oliver Aberth

Mathematics Department
Texas A & M University
College Station Texas

ACADEMIC PRESS
A Harcourt Science and Technology Company
San Diego San Francisco New York Boston
London Sydney Tokyo

This book is printed on acid-free paper. ∞

Academic Press
A Harcourt Science and Technology Company
525 B Street, Suite 1900, San Diego, California 92101-4495, USA
http://www.academicpress.com

Academic Press
Harcourt Place, 32 Jamestown Road, London NW1 7BY, UK
http://www.academicpress.com

Library of Congress Catalog Card Number: 00-111076

International Standard Book Number: 0-12-041752-9

PRINTED IN THE UNITED STATES OF AMERICA
01 02 03 04 05 06 IP 9 8 7 6 5 4 3 2 1

Dedicated
to the memory of
the definer of the computable numbers
Alan Mathison Turing
1912–1954

Contents

Preface

The concept of "finite means" arose in the 1930s when mathematicians dealt with a number of logical problems, and considered whether these problems could be solved in a constructive fashion. By a revolutionary way of reasoning, the Austrian Kurt Gödel obtained illuminating negative results, and he went on to show that the foundations of mathematics must always be incomplete. The "finite means" concept is useful in other areas of mathematics besides logic, and especially so in numerical analysis.

Consider the following programming task: We want to create a program that allows a user to define an arbitrary real number a, whereupon the program prints out a five-decimal place approximation to the number that is correct to the last decimal place. We will require the user to define the real number a unequivocally by giving us an algorithm for computing arbitrarily accurate rational approximations to a. Should we try to actually create such a program?

As simple as this programming task may appear, there is no program that can always obtain the correct decimal approximation, because this task cannot be done by "finite means." We may devise a program that obtains a correct five-decimal approximation for many, perhaps most, of the real number approximations entered, but there will always be some real number, algorithmically defined, for which our procedure fails. If we are aware of the "finite means" difficulty that lurks in this computation, we are warned that any program we devise is certain to have some loophole in it. Forewarned this way, we can try to amend the task in some way to escape this difficulty. If we require that the computed correct decimal place approximation be either five- or six-place, we find that this alternative task *can* be done by finite means.

Computable calculus can be useful when considering numerical programming tasks, because it can often give us either a red or a green light for the undertaking, that is, indicate whether or not the task can always be carried out by finite means. A red light does not mean that the task is impossible for any particular input, only that any conceivable general method of task execution must fail for some instances of the input.

In a previous book, *Precise Numerical Methods Using* C++ (published in 1998 by Academic Press), we described methods of obtaining precise numerical answers to many of the standard problems of elementary numerical analysis. That book analyzed computational problems according to computable calculus precepts, but only sketched the mathematical justification. In this book the mathematics is fleshed out.

The "finite means" way of analyzing numerical tasks is described in Chapter 3, where the mathematical reasoning is first only sketched. The reasoning is given in full detail in Chapter 5, where a certain ideal computer is defined to make concrete the notion of "finite means." When going through Chapter 5, the reader can use the accompanying software to understand the argument more easily, watching how the ideal computer can be programmed to do every computation we can do, and seeing why the computer must fail sometimes to correctly execute a certain simple computation task, that of deciding whether an arbitrary algorithmically defined real number is or is not equal to zero.

Acknowledgments

Through many years my colleague, G. Robert Blakley, discussed with me ways of improving the presentation of calculus, first at the University of Illinois in Urbana and later at Texas A&M University in College Station. His original ways of viewing calculus helped maintain my own interest in its foundations.

I am indebted to Ramon Moore of Ohio State University, not only for his interval arithmetic, which plays a central role in this book, but also for the constant encouragement he gave me while I was writing the text.

I thank my editor at Academic Press, Robert E. Ross, for his general helpfulness, but especially for suggesting the Windows format for the accompanying software. In addition, I thank Holly S. Bisso and Julie Bolduc, production editors at Academic Press, who cheerfully worked with me on the book.

The diagrams of this book were created using the `PicTex` program of Michael J. Wichura of the University of Chicago.

I could always turn to my wife Dawn for help in finding the right word in a phrase or sentence. Needless to say, this book would not have been possible without her.

Oliver Aberth

CHAPTER 1

Introduction

1.1 What Is Computable Calculus?

A simple way to introduce our subject is to relate it to another development in mathematics, the clarification of the concept of limit. In the early days of calculus, each student of the subject grasped the idea of a limit gradually, by working through various instructive examples. However, in complicated situations, even among experienced mathematicians, the individual interpretations of what a limit is sometimes led to disagreements as to whether a certain limit existed. The mathematician Augustin Cauchy dispelled this ambiguity about limits by providing the familiar epsilon—delta definition. Similarly, computable calculus gives an explicit meaning to the phrase "constructively possible," and in this way clarifies thinking about which computations are always feasible and which are not.

The founding work of the computable calculus point of view is a remarkable paper by Turing with the title "On Computable Numbers, with an Application to the Entscheidungsproblem." With the Entscheidungsproblem, there was the central idea of a *mechanical method* of solving certain logical questions. This was sometimes expressed by using phrases such as "by finite means" or "by constructive measures." This requires one to grapple with what it means to solve a mathematical question this way. The method Turing used to settle the Entscheidungsproblem was intuitively convincing that the "by finite means" aspect was correctly captured, because he proposed certain primitive machines, now known as Turing machines, and converted the Entscheidungsproblem into a question of whether these machines could solve it. Turing showed that for any machine that supposedly solved the problem, one could construct a specific problem for which the machine failed, and thus that the Entscheidungsproblem does not have a general constructive solution.

As the title of his paper indicates, Turing applied his machine concept to other parts of mathematics besides logic, and considered which numbers could be defined by his machines. These are the *computable numbers* of the title, the real numbers for which arbitrarily precise approximations can be computed by a machine. In this way Turing founded a different style of mathematics, which, using his adjective, we may call *computable mathematics* or *computable analysis*. Here all mathematical entities are required to be constructively defined or, to use Turing's approach, capable of being designated explicitly to an ideal computing machine for further manipulation and computation.

Actually, mathematicians who followed Turing made it clear that a computable form of mathematics was possible. The American, Hugh G. Rice, contributed to the subject, as did the English logician R. L. Goodstein and the Swiss mathematician Ernst Specker. However, it was a group of Russian mathematicians who extended Turing's computability ideas to such an extent that a new type of mathematical analysis was created. The founder of this school was A. A. Markov, the grandson of the well-known Markov of Markov processes, and his initial work was extended enormously by two other Russians, G. S. Ceitin and I. D. Zaslavsky. Later contributors were B. A. Kushner, who wrote a textbook on the subject [23], and the mathematicians V. P. Orevkov and O. Demuth. The *References* section in the back of this book lists key papers relevant to computable analysis by all the mathematicians mentioned. The development of a computable analysis as an alternative to conventional mathematical analysis was essentially complete by 1975, although today this analysis is largely unknown. It is the purpose of this book to make computable analysis more easily accessible, and toward that end we present a version of constructively defined calculus, or *computable calculus*. The computable analysis outlook can be used with any mathematical discipline, so there could be a computable complex analysis, or a computable topology.

It should be emphasized that it is never an issue as to which analysis is correct, that is, conventional analysis or computable analysis. Both ways of treating mathematical subjects are internally consistent, and results obtained using one analysis can be assigned an appropriate interpretation in the other analysis without difficulty. Some mathematical subjects, such as rational number theory, are treated essentially the same from either viewpoint. However, a real number is a subtle mathematical construct, and as soon as one uses real numbers in a mathematical discipline, differences can arise.

1.2 What Are the Advantages of Computable Calculus, If Any?

A perfectly natural reaction at this point is to ask "Why bother? Calculus has been in use now for over three centuries, and what possible reason is there for altering its rules?" The simplest answer to this question must be "to improve computation." Before the present age of high-speed computers, scientists and engineers had to be content with whatever approximate result they could laboriously extract with mechanical desk calculators. Now that we all have become accustomed to our powerful PCs and Macintoshes, and the millions of steps they can execute in a second, the problems we solve are becoming more extensive. In spite of this power of today's computers, however, we still do not solve our mathematical problems precisely, that is, to a specific number of correct decimal places.

Computable analysis gives insight into problems of computation, making it easier to generate correct answers on a computer. If we intend to do some computation, then we need to differentiate between a computation process that always takes a finite number of steps and a computation process that may take an infinite number of steps. We can attempt to execute the first type of process on an actual computer, but are well advised to avoid the other kind.

Computable calculus is calculus with the "by finite means" aspect made central. In order to ensure that our computations take "a finite number of steps," we develop in this book elementary calculus in constructive fashion. All the key mathematical concepts—the real numbers, sequences, functions, and so on—are defined in terms of some computation that an ideal computer could perform. Imagining an ideal computer is merely one approach among many proposed, for making exact the concept of *finitely defined process*. Three other standard ways of achieving this exactitude are through *recursive functions*, *lambda processes* or *normal algorithms*, each method extensively used by numbers of mathematicians. Besides these methods, there are numerous others less well known. All such methods for defining constructivity that are sufficiently general have turned out to be equivalent to each other, so the particular method used does not influence the conclusions reached. The ideal computer concept is appropriate for today's world where computers are everywhere, and a calculus student is expected to be comfortable in their use. An ideal computer is "ideal" in two senses: Its memory is finite but always large enough to accommodate whatever computation is contemplated. And no notice is taken of the number of steps the computer needs to carry out a task, as long as the number of steps is *finite*.

Our ideal computer is not completely specified until the middle of this book, because a detailed specification early on is not needed, and probably hinders the rapid development of computable calculus. For a treatment of this subject using a specific ideal computer different from the one proposed in this book, and where the ideal computer is spelled out from the start, a reader may consult the author's previous book [4]. Here we presume a reader's experience with practical programming enables the reader to follow the computability assertions. Real numbers are defined in ideal computer terms in the next chapter, the key concepts of solvable and nonsolvable problems appear in the chapter after that, sequences and functions are defined next, and finally in Chapter 5, our ideal computer is mapped out in detail.

To make the ideal computer concept more concrete, the accompanying CD contains software that can simulate the ideal computer on a PC. Directions for loading the CD software onto a PC's hard disk are given in the last page of this book. In the ideal computer chapter, there are simulation exercises suggested that will help clarify many of the computability assertions made in earlier chapters.

1.3 A Brief Description of Computable Calculus

In the next chapter we develop the theory of real numbers from the point of view of computable analysis. In that chapter some standard properties of the real numbers are derived and these properties are also obtained by a conventional treatment of the real numbers. Major differences between a conventional and a computable development first arise in Chapter 3, where it is shown that some simple operations with real numbers are not always "finitely executable."

Perhaps the main adjustment needed to become comfortable in computable calculus is the realization that deciding whether or not two real numbers are equal can sometimes turn into a difficult problem, that is, "take an infinite number of steps." This is a phenomenon that does not appear when dealing with rational numbers; for any two rational numbers p_1/q_1 and p_2/q_2, we can always decide whether or not they are equal.

This difficulty in deciding equality has the consequence that a function $f(x)$, which has a discontinuity at some x argument, cannot be a computable function. As an example, suppose we require $f(x)$ to have the value 0 for $x < \frac{1}{3}$, and to have the value 1 for $x \geq \frac{1}{3}$, so that $f(x)$ has a jump discontinuity at $x = \frac{1}{3}$ (see Fig. 1.1). Now imagine that the x arguments we receive are computable and can be approximated arbitrarily accurately. That is, we can obtain a decimal approximation accurate to as many decimal places as we please, except for the last decimal place obtained, and this last decimal is off

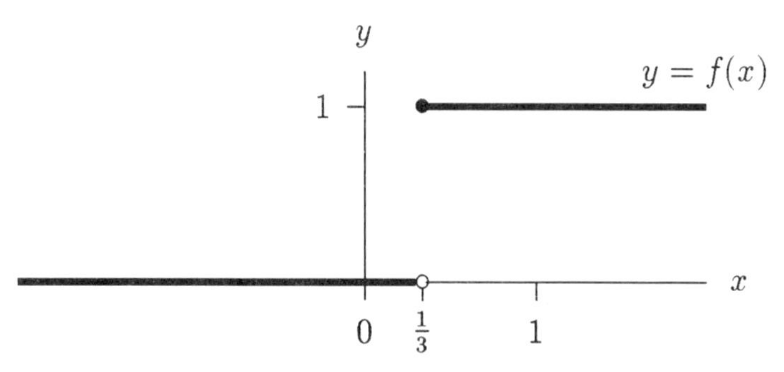

Figure 1.1. A certain function $f(x)$.

only by at most 1. We can supply the proper $f(x)$ value for most x arguments, but there is one case that could cause us trouble. Suppose we obtain a succession of more and more accurate x approximations and they all have the same form: $0.3333\cdots333^{\pm}1$. In this case we are at a loss as to which $f(x)$ value to assign, 0 or 1. We cannot continue getting better and better approximations indefinitely, because the situation may never change, x being equal to $\frac{1}{3}$. But if we assign the value 1 after examining any number of such approximations, this assigned value may be incorrect, because an x approximation more accurate than any obtained so far might be $0.3333\cdots331^{\pm}1$, so the $f(x)$ value we should have assigned is 0 and not 1.

In order to use in computable calculus the function $f(x)$ of Fig. 1.1, we first must grant that with respect to actual computation this function is really *undefined* at $x = \frac{1}{3}$. Once we grant this, and give up trying to compute the function's value at $x = \frac{1}{3}$, the function becomes computable, its definition now being 0 if $x < \frac{1}{3}$, 1 if $x > \frac{1}{3}$, and *undefined* at $x = \frac{1}{3}$. In the Fig. 1.1 diagram, the small dark circle to the right of the y-axis 1 mark would need to be replaced by an empty circle, like the one at the x-axis $\frac{1}{3}$ mark.

The consequences of not being able to have discontinuous functions in computable calculus go to the core of the subject. The derivative $f'(x)$ of a function $f(x)$ has to be defined slightly differently in order to avoid assigning discontinuous derivatives to continuous functions. In addition, the derivative change requires an adjustment in the Cauchy definition of limits, because derivatives get defined via limits.

Of course many traditional parts of calculus remain essentially unchanged in computable calculus. For instance, the theory of the Riemann integral, itself beautifully constructive, is easily accommodated in computable calculus. (This would not be the case with the Lebesgue integral, which is not a constructive integral.)

In a few cases computable calculus does not obtain certain standard results of conventional calculus. This is because these standard results imply some computation not achievable "in a finite number of steps." There are four theorems in this book that are not needed for the development of the subject, but are present to identify areas of divergence with conventional calculus. The first of these theorems appears in Chapter 6 and is Specker's theorem, which gives an example of a bounded monotone sequence that does not converge to a limit. Another Chapter 6 result diverging from conventional calculus is Theorem 6.9, giving an example of a function continuous at every point in a finite closed interval, but unbounded there and consequently not uniformly continuous over the interval. Theorem 7.4 has an example of a function uniformly continuous over a finite closed interval, the function not assuming a maximum value at any point in that interval. The last result diverging from conventional calculus is Theorem 11.2, giving an example of the initial value problem of differential equations having no solution, whereas Peano's theorem of conventional calculus implies that a solution exists.

These four instructive results easily have the most intricate constructions of this book. It may appear from them that computable analysis requires more complicated reasoning than conventional analysis. To dispel this feeling, let us imagine a freshman level computable calculus text. Such a text, not making the assertions that have been countered by the four theorems listed, would not need to delve into the intricate constructions of these theorems. The calculus presentation in such a text would probably be comparable in complexity to that of a conventional calculus text, being much the same in many respects, but of course having some key differences. The initial chapter of such a text would need a thorough treatment of the topic of accurate computation, so that the difficulty of computing the value of a function at a point of discontinuity is made clear.

CHAPTER 

The Real Numbers

2.1 Definition of a Real Number

Each mathematical concept beyond integers and rational numbers requires a definition expressed as some computation that can be carried out by an ideal computer. Here, following Turing, we are using the concept of an ideal computer as our means of making all definitions explicitly constructive. We use the adjective "computable" to mean "executable by an ideal computer in a finite number of steps." The ideal computer certainly can do integer computations, and the integers can be arbitrarily large. It can also do rational number computations, because a rational number may be viewed as defined by an ordered pair of integers. The problem now is how the ideal computer will deal with real numbers.

There have been several definitions of real numbers that have been used in mathematics, and each of these can be changed to a computable form. For our definition of a real number, we follow the Weierstrass concept of a real number and suppose we have some method of obtaining arbitrarily accurate rational approximations to the number.

We allow a real number to be defined in three distinct but equivalent ways. A real number a is defined by a computable function $a(E)$ of the positive rational E, with $a(E)$ equal to $m \pm e$, where m is a rational approximant to a, the number e is a nonnegative rational error bound for m that is not greater than E, and the number a is guaranteed to lie in the interval $m \pm e$ (see Fig. 2.1). This interval can also be expressed as $[m - e, m + e]$, using the endpoint notation.

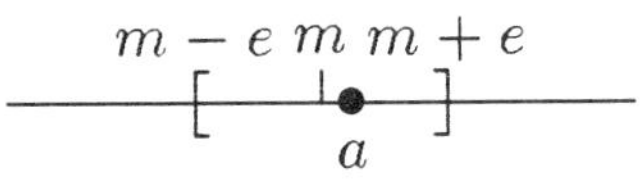

Figure 2.1. The interval $m \pm e$.

As an example, the number π is defined by a function $\pi(E)$ which might supply the following approximations:

$$\pi(E)$$

$$\pi\left(\frac{1}{10}\right) = \frac{505}{162} \pm \frac{55}{1944}$$

$$\pi\left(\frac{1}{100}\right) = \frac{6115}{1944} \pm \frac{2315}{489888}$$

Notice how difficult it is to tell whether the approximant $\frac{505}{162}$ is as accurate as it is supposed to be. We are more accustomed to decimal approximations to real numbers rather than rational ones. As a second option, then, we allow the approximating function to give its approximants in decimal form with the approximant always having an error bound of one unit in the approximant's last decimal place. We distinguish this second method of approximation by using the argument letter D to denote the number of decimal places wanted, so D is a positive integer this time instead of a positive rational as was E. For instance, continuing with the π example, our function $\pi(D)$ might supply the values

$$\pi(D)$$

$$\pi(2) = 3.15^{\pm}1$$

$$\pi(3) = 3.142^{\pm}1$$

$$\pi(4) = 3.1416^{\pm}1$$

Here $3.142^{\pm}1$ denotes the interval $[3.141, 3.143]$ with endpoints obtained by either subtracting 1 from the last decimal digit of the approximant 3.142 or by adding 1 to this last digit. The second type of approximation function $a(D)$ can always be obtained constructively from the first type of approximation function $a(E)$ in this way: Given D, the number of decimal places wanted, we set E equal to half the D error bound, that is, $\frac{1}{2}10^{-D}$, and then $a(E)$ delivers a rational approximant p/q with error bound not greater than $\frac{1}{2}10^{-D}$. Using the ordinary division algorithm, we divide q into p and obtain a D decimal digit quotient $d_0.d_1d_2 \cdots d_D$, with the error of the division being <1 unit in the last (Dth) place, or 10^{-D}. To reduce the error bound of this division to a half unit in the last place, we obtain an extra quotient digit d_{D+1}. If this extra digit is 5, 6, 7, 8 or 9, we add 1 to the last decimal place of our approximant, otherwise we leave the approximant unchanged. The D decimal approximant now can be assigned an error bound of 1 unit in its last decimal place, this error bound being the sum of the E error bound and the division error bound.

For instance, to compute $\pi(D)$ with D equal to 2, we take E equal to $\frac{1}{2}10^{-2}$ and then obtain the approximant of $\pi(E)$, which turns out to be 6115/1944. We divide 1944 into 6115 to obtain a three-decimal place result, 3.145. The last decimal place is the digit 5, so we round up 3.14 to 3.15, and our $\pi(D)$ value is $3.15^{\pm}1$.

For easier proofs of various properties of real numbers, it is convenient to employ one final type of approximation function, and we distinquish this third type of function by using the argument letter K, which like D denotes a positive integer. This approximation function $a(K)$ gives an output like the function $a(E)$ in that a rational approximant m_K and an error bound e_K are supplied, with the error of the approximant guaranteed to be no greater than e_K. It is convenient to allow the error bound e_K even to take on an infinite value (designated by the symbol ∞) for a finite number of K arguments. As K increases, e_K in general decreases, but it is best not to be too restrictive in the way that e_K decreases with K. We require only that if a positive rational error bound E is prescribed, an integer K_0 exists such that $e_K < E$ for $K > K_0$. A computable way of obtaining K_0 as a function of E is not required, only the certainty that for every E an integer K_0 exists.

Given a function $a(D)$, we can obtain a function $a(K)$ by converting the decimal approximant of $a(D)$ and the error bound to correct rational form. For instance, continuing with the π example, the approximating function $\pi(D)$ can be converted to a function $\pi(K)$, which supplies these values:

$$\pi(K)$$

$$\pi(2) = \frac{315}{100} \pm \frac{1}{100}$$

$$\pi(3) = \frac{3142}{1000} \pm \frac{1}{1000}$$

$$\pi(4) = \frac{31416}{10000} \pm \frac{1}{10000}$$

Suppose for a real number a we have the approximation function $a(K)$ of this third variety, and we want the more convenient approximation function $a(E)$. It is possible to obtain $a(E)$ constructively from $a(K)$ because we can compute $a(K)$ for larger and larger values of K, and eventually must find an interval $m_K \pm e_K$ with $e_K \leq E$.

Let us spell out the requirements for a real number using this last type of approximation function.

TENTATIVE DEFINITION 2.1: *A real number a is defined by a program $a(K)$ that for any value of the positive integer K supplies a rational approximant m_K and a nonnegative rational error bound e_K defining an interval $m_K \pm e_K$ containing the real number a, that is, with the inequality $|m_K - a| \leq e_K$ true. Given any positive rational error bound E, it is certain that an integer K_0 exists such that $e_K < E$ for $K > K_0$.*

A flaw of our definition of a real number a is the specification $|m_K - a| \leq e_K$, because the real number a being defined gets involved in its own definition. We need to avoid this circularity and, to this end, we introduce some new notation for the comparison of intervals.

2.2 An Ordering of Intervals

The intervals we are considering are intervals of rational numbers and these intervals have rational endpoints. Any two such intervals either intersect each other, or one interval is to the left of the other. With intervals we use the dotted symbols $\lessdot$, $\doteq$, $\gtrdot$ to indicate order relations. For any two intervals $m_1 \pm e_1$ and $m_2 \pm e_2$, we write

$$m_1 \pm e_1 \lessdot m_2 \pm e_2$$

if the left endpoint of the second interval is greater than the right endpoint of the first interval, that is, if $m_2 - e_2 > m_1 + e_1$. If we have neither $m_1 \pm e_1 \lessdot m_2 \pm e_2$ nor $m_2 \pm e_2 \lessdot m_1 \pm e_1$, that is, the two intervals intersect, we write

$$m_1 \pm e_1 \doteq m_2 \pm e_2$$

This occurs if and only if

$$\begin{aligned} m_2 - e_2 \leq m_1 + e_1 \quad &\text{and} \quad m_1 - e_1 \leq m_2 + e_2 \\ m_2 - m_1 \leq e_1 + e_2 \quad &\text{and} \quad m_1 - m_2 \leq e_1 + e_2 \\ \max(m_2 - m_1, m_1 - m_2) &\leq e_1 + e_2 \\ |m_1 - m_2| &\leq e_1 + e_2 \end{aligned} \tag{2.1}$$

The dotted symbols $\doteq$, $\gtrdot$ and $\lessdot$ signal that intervals are being compared. There is no difficulty in carrying out this comparison of two intervals because only rational interval endpoints are compared. It is always true for any two intervals $m_1 \pm e_1$ and $m_2 \pm e_2$, that exactly one of the three order relations hold:

$$m_1 \pm e_1 \lessdot m_2 \pm e_2, \quad m_1 \pm e_1 \doteq m_2 \pm e_2, \quad \text{or} \quad m_1 \pm e_1 \gtrdot m_2 \pm e_2$$

The other order symbols $\geq$, $\leq$ and $\neq$ all may be given a dotted form:

$\dot{\geq}$, $\dot{\leq}$ and $\dot{\neq}$, with the obvious interpretations. Thus $m_1 \pm e_1 \dot{\neq} m_2 \pm e_2$ means $m_1 \pm e_1 \dot{<} m_2 \pm e_2$ or $m_1 \pm e_1 \dot{>} m_2 \pm e_2$.

The dotted symbols may also be used to compare an interval with a rational number r, because we may interpret r as a point interval $r \pm 0$. For instance, we may occasionally write $m \pm e \dot{>} 0$ to indicate that the interval $m \pm e$ is to the right of the zero point. Such an interval we call *positive*. Similarly, an interval $m \pm e$ is *negative* if $m \pm e \dot{<} 0$.

Returning to the problem of eliminating the circularity from our first definition, we now have:

FINAL DEFINITION 2.1: *A real number a is defined by a program $a(K)$ that for any value of the positive integer K supplies a rational approximant m_K and a nonnegative rational error bound e_K defining an interval $m_K \pm e_K$ of rational numbers. For any choice of integers K_1 and K_2, the intervals $a(K_1)$ and $a(K_2)$ intersect, that is, we have*

$$a(K_1) = m_{K_1} \pm e_{K_1} \doteq m_{K_2} \pm e_{K_2} = a(K_2)$$

Given any positive rational error bound E, it is certain that an integer K_0 exists such that $e_K < E$ for $K > K_0$.

Now the real number a is not mentioned in its definition. The number a is defined by the computable algorithm $a(K)$ specifying an infinite set of intersecting intervals of *rational numbers*, so now the definition of a real number is purely in terms of the rationals.

We emphasize that a real number is specified by giving its *ideal computer program*. This means that a real number is defined not by any finite list of approximation values, but by an ideal computer program that produces the approximations. By changing the steps of $a(K)$ in various ways, perhaps even adding steps that accomplish nothing, from any given program $a(K)$ we can generate an infinite number of alternate ideal computer programs supplying acceptable values. The situation here is similar to the situation with the field of rational numbers. For any rational number $r = p/q$, there are an infinite number of ordered integer pairs p, q that can be used to define the rational, all equally suitable to specify the rational. For a real number a, there are an infinite number of ideal computer programs to specify the real number.

Definition 2.1 explicitly specifies the approximation function $a(K)$. However, if an approximation function $a(E)$ or $a(D)$ were supplied instead, this would be acceptable. We have shown that a function $a(E)$ can be constructively obtained from a function $a(K)$, that a function $a(D)$ can be constructively obtained from a function $a(E)$, and that a function $a(K)$ can be constructively obtained from a function $a(D)$. Thus the three forms are equivalent to each other and any one of the three can be used to define a real number. In later sections we may employ the functions $a(E)$ or $a(D)$ if that becomes convenient. An $a(E)$ function must be such that for any E_1 and E_2 the interval $a(E_1)$ intersects the interval $a(E_2)$, in order to ensure that a derived $a(K)$ function satisfies the intersection requirement of Definition 2.1. Similarly, an $a(D)$ function must be such that for any D_1 and D_2 the interval $a(D_1)$ intersects the interval $a(D_2)$.

The functions $a(K)$, $a(E)$, or $a(D)$ play a central role in computable calculus, and accordingly we distinguish these functions by calling them *approximation algorithms*. From now on we use this term for them exclusively.

For each rational number p/q there is a corresponding real number defined by the approximation algorithm $(p/q)(K)$ giving the interval $p/q \pm 0$ for all K. As we will see shortly, real number arithmetic operations are defined in such a way that the real numbers defined by algorithms $(p/q)(K)$ can be combined by operations $+$, $-$, $\times$, $\div$ to mimic exactly the way rational numbers p/q combine under such operations in the rational field. There is no reason to use different notation for the rational p/q and the real number defined by $(p/q)(K)$. Whenever we designate some rational p/q as a real number, this is to be interpreted as meaning the real number with approximation algorithm $(p/q)(K)$.

Next we need to define when two real numbers are equal. The obvious definition is:

DEFINITION 2.2: *Two real numbers a and b, defined by approximation algorithms $a(K)$ and $b(K)$, respectively, are equal if and only if*

$$a(K_1) \doteq b(K_2) \quad \textit{for all } K_1, K_2$$

That is, each of the infinitely many intervals defined by $a(K)$ intersects each of the infinitely many intervals defined by $b(K)$.

Our equality relation is reflexive ($a = a$), symmetric ($a = b$ implies $b = a$), and transitive ($a = b$ and $b = c$ implies $a = c$). Transitivity may be proved as follows: For any positive integers K_1 and K_2, and for any choice of the positive integer K, we have $a(K_1) \doteq b(K)$ and $b(K) \doteq c(K_2)$. Therefore the two intervals $a(K_1)$ and $c(K_2)$ both intersect any interval $b(K)$. We need to show that $a(K_1)$ and $c(K_2)$ intersect each other. However, if they failed to intersect each other, we would have a contradiction. If δ is the gap between these two intervals, we would be able to find a K such that the length of the interval $b(K)$ was $< \delta$, and then $a(K_1)$ and $c(K_2)$ could not both intersect $b(K)$.

Any reflexive, symmetric and transitive relation between objects divides the objects into equivalence classes. Here each equivalence class of approximation algorithms may be considered as associated with a particular real number.

If the numbers a and b are equal according to our definition, we write $a = b$, and if they are not equal, that is, there are integers K_1 and K_2 such that $a(K_1) \not\doteq b(K_2)$, then we write $a \neq b$. We can be more specific: We write $a > b$ when $a(K_1) \gtrdot b(K_2)$ and $a < b$ when $a(K_1) \lessdot b(K_2)$. It is not possible that we have both $a < b$ and $a > b$ because this implies that we can choose K_1, K_2, K_1', K_2' to obtain both $a(K_1) \lessdot b(K_2)$ and $a(K_1') \gtrdot b(K_2')$. Because the intervals $a(K_1)$ and $a(K_1')$ intersect, we can find a rational r_a common to both intervals. Similarly, we can find a rational r_b common to both intervals $b(K_2)$ and $b(K_2')$. This gives us a contradiction because then we obtain both $r_a < r_b$ and $r_a > r_b$.

The inequality relation is consistent for all a, b algorithms. That is, if we find $a(K_1) \lessdot b(K_2)$, and $a'(K)$ and $b'(K)$ are other approximations algorithms for the numbers a and b, then we can also find integers K_1' and K_2' such that $a'(K_1') \lessdot b'(K_2')$. If δ is the gap between $a(K_1)$ and $b(K_2)$, we need only choose K_1' and K_2' large enough so that the lengths of the intervals $a'(K_1')$ and $b'(K_2')$ are both $< \delta/2$.

We see now that for any two real numbers a and b, we have exactly one of the three possibilities: $a = b$, $a < b$, or $a > b$. As usual, we define $a \leq b$ to mean $a = b$ or $a < b$. Notice that if we have $a(K) = m \pm e$, it follows that

$$m - e \leq a \leq m + e$$

The inequality $m - e \leq a$ must be true because we obtain a contradiction if we assume the opposite, $m - e > a$. This assumption implies the existence of integers K_1 and K_2 such that the interval $a(K_1)$ is to the left of the interval $(m - e)(K_2) = (m - e) \pm 0$, contradicting the interval intersection relation $a(K_1) \doteq a(K) = m \pm e$. The inequality $a \leq m + e$ is true by similar reasoning. Then a real number a defined by an approximation algorithm $a(K)$ must lie in every interval $a(K)$, and if a is defined by an approximation algorithm $a(D)$ or $a(E)$, the same reasoning shows that a lies in each interval $a(D)$ or $a(E)$.

With equality, inequality and order relations for the real numbers defined, the next step is to show that the real numbers form a field, that is, satisfy the following field axioms:

(i) Commutative laws	$a + b = b + a$	$ab = ba$
(ii) Associative laws	$a + (b + c) = (a + b) + c$	$a(bc) = (ab)c$
(iii) Distributive law	$a(b + c) = ab + ac$	
(iv) Identity elements	$a + 0 = a$	$a1 = a$
(v) Existence of inverses	$\forall a \exists b$ such that $a + b = 0$	$\forall a \neq 0 \exists c$ such that $ac = 1$

Because the rational numbers are a field, they satisfy these axioms. Further, it is natural to expect that the real numbers inherit all these relations.

First we must decide what is the sum $a + b$ and what is the product ab of two real numbers a and b. Before doing this, it is convenient to digress a little and introduce the fundamentals of interval arithmetic. Interval arithmetic was invented in 1962 by the mathematician Ramon Moore [27]. This arithmetic has found so many applications that the name "interval arithmetic" now also denotes a useful mathematical discipline [6; 28; 29; 30].

2.3 Interval Arithmetic

Here we consider combining arithmetically two real number approximations, $m_1 \pm e_1$ and $m_2 \pm e_2$. The operation $(m_1 \pm e_1) \circ (m_2 \pm e_2)$, where $\circ$ can be any of the four rational operations $+, -, \times$, or $\div$, yields the approximant $m = m_1 \circ m_2$, but what error bound e should we ascribe to m? That is, we are supposing that

$$m_1 \pm e_1 \circ m_2 \pm e_2 = m \pm e$$

We take e to be just large enough so that if r_1 is any rational lying in $m_1 \pm e_1$ and r_2 is any rational lying in $m_2 \pm e_2$, then $r_1 \circ r_2$ lies in $m \pm e$. The relations we may use here are:

$$m_1 \pm e_1 + m_2 \pm e_2 = (m_1 + m_2) \pm (e_1 + e_2) \tag{2.2}$$

$$m_1 \pm e_1 - m_2 \pm e_2 = (m_1 - m_2) \pm (e_1 + e_2) \tag{2.3}$$

$$m_1 \pm e_1 \times m_2 \pm e_2 = (m_1 m_2) \pm (e_1|m_2| + |m_1|e_2 + e_1 e_2) \tag{2.4}$$

$$m_1 \pm e_1 \div m_2 \pm e_2 = \begin{cases} \left(\dfrac{m_1}{m_2}\right) \pm \left(\dfrac{e_1 + |\frac{m_1}{m_2}|e_2}{|m_2| - e_2}\right) & \text{if } |m_2| > e_2 \\ \text{Division error} & \text{if } |m_2| \le e_2 \end{cases} \tag{2.5}$$

To show that these error bounds are correct, we must show that for $\circ$ equal to any of the four operations, and for ε_1 and ε_2 any rationals satisfying the relations $|\varepsilon_1| \le e_1$ and $|\varepsilon_2| \le e_2$, or, equivalently, $-e_1 \le \varepsilon_1 \le e_1$ and $-e_2 \le \varepsilon_2 \le e_2$, the quantity $|(m_1 + \varepsilon_1) \circ (m_2 + \varepsilon_2) - (m_1 \circ m_2)|$ is less than or equal to the error bound displayed in the preceding, with equality holding for certain choices of ε_1 and ε_2. This implies that the $\circ$ result interval $m \pm e$ contains all rationals $r_1 \circ r_2$, where r_1 is any rational in $m_1 \pm e_1$ and r_2 is any rational in $m_2 \pm e_2$. For the four cases $+, -, \times$, and $\div$, we have:

$$|(m_1 + \varepsilon_1) + (m_2 + \varepsilon_2) - (m_1 + m_2)| = |\varepsilon_1 + \varepsilon_2| \le |\varepsilon_1| + |\varepsilon_2| \le e_1 + e_2$$

with equality if $\varepsilon_1 = e_1$ and $\varepsilon_2 = e_2$ or $\varepsilon_1 = -e_1$ and $\varepsilon_2 = -e_2$.

$$|(m_1 + \varepsilon_1) - (m_2 + \varepsilon_2) - (m_1 - m_2)| = |\varepsilon_1 - \varepsilon_2| \le |\varepsilon_1| + |\varepsilon_2| \le e_1 + e_2$$

with equality if $\varepsilon_1 = e_1$ and $\varepsilon_2 = -e_2$ or $\varepsilon_1 = -e_1$ and $\varepsilon_2 = e_2$.

$$
\begin{aligned}
|(m_1 + \varepsilon_1) \times (m_2 + \varepsilon_2) - (m_1 m_2)| &= |\varepsilon_1 m_2 + m_1 \varepsilon_2 + \varepsilon_1 \varepsilon_2| \\
&\leq |\varepsilon_1||m_2| + |m_1||\varepsilon_2| + |\varepsilon_1||\varepsilon_2| \\
&\leq e_1 |m_2| + |m_1| e_2 + e_1 e_2
\end{aligned}
$$

with equality if $\varepsilon_1 = (m_1 \text{ sign}) e_1$ and $\varepsilon_2 = (m_2 \text{ sign}) e_2$, because this gives the three quantities $\varepsilon_1 m_2$, $m_1 \varepsilon_2$, $\varepsilon_1 \varepsilon_2$ the same sign. (If $m_1 = 0$, then (m_1 sign) may be taken $+$ or $-$.) (If $m_2 = 0$, then (m_2 sign) may be taken $+$ or $-$.)

$$
\begin{aligned}
\left| \frac{m_1 + \varepsilon_1}{m_2 + \varepsilon_2} - \frac{m_1}{m_2} \right| &= \frac{|\varepsilon_1 m_2 - m_1 \varepsilon_2|}{|(m_2 + \varepsilon_2) m_2|} \\
&\leq \frac{|\varepsilon_1||m_2| + |m_1||\varepsilon_2|}{|m_2 + \varepsilon_2||m_2|} = \frac{|\varepsilon_1| + \frac{|m_1|}{|m_2|}|\varepsilon_2|}{|m_2 + \varepsilon_2|} \\
&\leq \frac{e_1 + \frac{|m_1|}{|m_2|} e_2}{|m_2 - e_2|}
\end{aligned}
$$

with equality if $\varepsilon_1 = (m_1 \text{ sign}) e_1$ and $\varepsilon_2 = (m_2 \text{ sign}) e_2 \cdot (-1)$, because this gives the two quantities $\varepsilon_1 m_2$ and $-m_1 \varepsilon_2$ the same sign. (If $m_1 = 0$, then (m_1 sign) may be taken $+$ or $-$.)

These relations show that the error bound given for an operation $\circ$ is neither too small nor too large.

Note that the interval relations also show how to properly multiply an interval by a rational number, for instance, what the product $2 \times (3 \pm 2)$ is. For the operation $r \times (m \pm e)$ or the operation $(m \pm e) \times r$, we simply convert the rational r to a point interval and use the fore-mentioned multiplication relation. That is, r is thought of as $r \pm 0$, so $r \times (m \pm e)$ equals $(r \pm 0) \times (m \pm e)$. Thus, using Eq. (2.4), we have

$$2 \times (3 \pm 2) = (2 \pm 0) \times (3 \pm 2) = 6 \pm (0 \cdot 3 + 2 \cdot 2 + 0 \cdot 2) = 6 \pm 4$$

For that matter, the four interval arithmetic relations give the correct answer when *both* operands are exact rationals, because an exact result of the form $m \pm 0$, that is, a point, is obtained.

After we define how to execute the four rational operations with real numbers we can verify the field axioms.

2.4 The Real Number Operations $+, -, \times, \div$

Given the real numbers a and b defined by approximation algorithms $a(K)$ and $b(K)$, and a rational operation $\circ$ chosen from the possibilities $+, -, \times, \div$, we must specify a suitable approximation algorithm for $a \circ b$. Our definition is in terms of interval arithmetic and is:

$$(a \circ b)(K) = a(K) \circ b(K)$$

A K approximation algorithm is allowed to have a finite number of arguments K which lead to an approximant with an infinite error bound, and if this case should occur for either $a(K)$ or $b(K)$, then instead of obtaining $(a \circ b)$ by interval arithmetic, we simply compute the approxminant in the standard way and assign an infinite error bound. This convention can also be thought of as a slight extension of the interval arithmetic equations so that whenever $m_1 \circ m_2$ is defined, we have

$$(m_1 \pm \infty) \circ (m_2 \pm e_2) = (m_1 \circ m_2 \pm \infty)$$

and

$$(m_1 \pm e_1) \circ (m_2 \pm \infty) = (m_1 \circ m_2 \pm \infty)$$

Consider first the operations of addition, subtraction, and multiplication. Our algorithm $(a \circ b)(K)$ is computable if the algorithms $a(K)$ and $b(K)$ are. Also, given any rational error bound E, an integer K_0' exists such that for $K > K_0'$ the error bound of $a(K)$ is $< E$. Similarly, an integer K_0'' exists such that for $K > K_0''$ the error bound of $b(K)$ is $< E$. This implies that an integer K_0''' exists such that for $K > K_0'''$ *both* approximations $a(K)$ and $b(K)$ have error bounds $< E$. Therefore the formulas of interval arithmetic are certain to be used for $K > K_0'''$, and these imply that $(a \circ b)(K)$ has the required K algorithm property that for any prescribed E, an integer K_0 exists such that for $K > K_0$ the error bound of $(a \circ b)(K)$ is $< E$.

For division, we must take into account that the interval division operation does not yield a result interval if the divisor interval includes the zero point. However, the algorithm $(a \div b)(K)$ for $b \neq 0$ is always defined if K is taken large enough, as we may see by the following argument. According to Definition 2.2, if $b \neq 0$, there is a K_1 and a K_2 such that $b(K_1) = m_{K_1} \pm e_{K_1}$ does not intersect $(0/1)(K_2) = 0 \pm 0$ (the zero point). This implies $|m_{K_1}| > e_{K_1}$. If δ equals $|m_{K_1}| - e_{K_1}$, there is an integer K_0 such that for $K > K_0$ the $b(K)$ interval length is less than δ, implying that the interval $b(K)$ cannot contain the zero point because $b(K)$ intersects $b(K_1)$. Thus if $b \neq 0$, then an integer K_0 exists such that for $K > K_0$ we have $(a \div b)(K)$ defined as a finite interval. The conclusion that $(a \div b)(K)$ is computable and decreases in the way required now follows in the same way as for addition, subtraction, and multiplication.

To make the description for the operation $a \div b$ more like the other cases of $a \circ b$, it is convenient to revise the definition of the interval arithmetic division operation to

$$m_1 \pm e_1 \div m_2 \pm e_2 = \begin{cases} \left(\dfrac{m_1}{m_2}\right) \pm \left(\dfrac{e_1 + |\frac{m_1}{m_2}|e_2}{|m_2| - e_2}\right) & \text{if } |m_2| > e_2 \\ 0 \pm \infty & \text{if } |m_2| \le e_2 \end{cases}$$

When $|m_2| \le e_2$, the result $0 \pm \infty$ defines an interval equal to the whole real line. (The 0 approximant is an arbitrary choice and has no significance.)

It still remains to be shown that for any K_1 and K_2 we have

$$(a \circ b)(K_1) \doteq (a \circ b)(K_2)$$

This result follows easily from the properties of interval arithmetic. The interval $(a \circ b)(K)$ contains every rational $r_a \circ r_b$, where r_a is any rational in $a(K)$ and r_b is any rational in $b(K)$. We have $a(K_1) \doteq a(K_2)$, so there is some rational r'_a that lies in both intervals $a(K_1)$ and $a(K_2)$. Similarly, $b(K_2) \doteq b(K_2)$, so there is some rational r'_b that lies in both intervals $b(K_1)$ and $b(K_2)$. Then $r'_a \circ r'_b$ lies in both intervals $(a \circ b)(K_1)$ and $(a \circ b)(K_2)$, so these intervals intersect.

A remaining concern at this point might be this: Although the method proposed for defining the four rational operations delivers a computable approximation algorithm $(a \circ b)$ from the computable approximation algorithms $a(K)$ and $b(K)$, if we have $a = a_1$ and $b = b_1$, is the real number defined by $(a \circ b)(K)$ equal to the real number defined by $(a_1 \circ b_1)(K)$? This is a question of the *consistency* of the rational operations. To put this concern another way, with several approximation algorithms to choose from for a real number a, and several approximation algorithms for another real number b, will the various approximation algorithms obtained for a rational operation $a \circ b$ always define the same real number? We need to show that $a \circ b$ equals $a_1 \circ b_1$, so we need to show that for any K_1 and K_2 we have

$$(a \circ b)(K_1) \doteq (a_1 \circ b_1)(K_2)$$

A proof of this relationship is like the proof of the previous paragraph.

Now we are able to consider the proof of the field axioms for real numbers. The additive identity element is, of course, the real number 0 with defining algorithm $(0/1)(K)$, and the multiplicative identity element is, of course, the real number 1 with defining algorithm $(1/1)(K)$. The additive inverse of a is $-a$ or $-1 \cdot a$, and if $a \neq 0$, its multiplicative inverse is $1/a$. The proof of any of the field axioms is a routine matter after we show:

LEMMA: *If two real numbers a and b with approximation algorithms $a(K) = m_K \pm e_K$ and $b(K) = m'_K \pm e'_K$ are such that for all K larger than some fixed integer $\widehat{K}$, we have $m_K = m'_K$ ($m_K \le m'_K$), then $a = b$ ($a \le b$).*

The lemma has two parts according to whether we have either $m_K = m'_K$ or $m_K \leq m'_K$. We prove only the case $m_K = m'_K$; the other case is shown similarly. We suppose $a \neq b$ and show that this leads to a contradiction. If $a \neq b$, there are positive integers K_1 and K_2 such that the interval $a(K_1)$ does not intersect the interval $b(K_2)$. Let δ equal the distance separating the two intervals. As we compute $a(K)$ and $b(K)$ for larger and larger $K > \widehat{K}$, we are certain to find a K' such that the error bounds for $a(K')$ and $b(K')$ are both $< \delta/4$. The interval $a(K')$ intersects the interval $a(K_1)$, and the interval $b(K')$ intersects the interval $b(K_2)$, but the interval $a(K')$ cannot intersect the interval $b(K')$ because the total length of these two intervals is $< \delta$. However these two intervals both contain identical approximants and so must intersect, and this gives us our contradiction.

Each of the field axioms may be proved in the same way, and as an example we consider only the distributive law, $a(b+c) = ab + ac$. A real number $ab + ac$ is arrived at after three rational operations $\circ$, and a real number $a(b+c)$ also is arrived at after three such operations. These two real numbers have K approximation algorithms that deliver intervals according to interval arithmetic operations using the intervals $a(K)$, $b(K)$, and $c(K)$. The various combinations of interval arithmetic operations do not necessarily deliver identical error bounds, but they do deliver identical approximants, so according to the lemma just shown, the real number $a(b+c)$ is equal to the real number $ab + ac$.

2.5 The Absolute Value of a Real Number

For a rational number r, its absolute value $|r|$ is defined by the equation

$$|r| = \begin{cases} r & \text{if } r \geq 0 \\ -r & \text{if } r < 0 \end{cases}$$

From this definition we obtain the following relations for rational numbers:

$$\begin{aligned} |r| &\geq 0 \\ -|r| &\leq r \leq |r| \\ |r_1 + r_2| &\leq |r_1| + |r_2| \\ |r_1 r_2| &= |r_1|\,|r_2| \\ |r_1 - r_2| &\geq \Big||r_1| - |r_2|\Big| \end{aligned} \tag{2.6}$$

We have defined the four rational operations on real numbers and now need to define the operation of taking the absolute value of a real number. We defined a rational operation $\circ$ of two real numbers by using the interval arithmetic operation for $\circ$. Similarly, to define the absolute value of a real number, it is convenient to first define the interval arithmetic operation of taking an absolute value:

$$|m \pm e| = |m| \pm e \tag{2.7}$$

For a real number a defined by the approximation algorithm $a(K) = m_K \pm e_K$, the K_1 interval intersects the K_2 interval, which by Eq. (2.1) is equivalent to

$$|m_{K_1} - m_{K_2}| \leq e_{K_1} + e_{K_2}$$

The real number $|a|$ is defined to be the real number with the approximation algorithm $|a|(K)$ equal to $|a(K)|$, which is $|m_K| \pm e_K$. By the relation displayed in the preceding, and the last relation of Eq. (2.6), we have

$$\Big| |m_{K_1}| - |m_{K_2}| \Big| \leq |m_{K_1} - m_{K_2}| \leq e_{K_1} + e_{K_2}$$

Making use of Eq. (2.1) once more, this implies that the $|a|(K)$ intervals intersect each other, as required for a real number approximation algorithm.

With the help of the preceding lemma, it is easy to see that all the relations given here for rational numbers are also true for real numbers.

CHAPTER 3

Solvable Problems and Nonsolvable Problems

3.1 Introduction

First we give Turing's proof that a certain problem, known as the halting problem, cannot be solved by finite means. After this we can show that various common problems of computation are similar to attempting to solve the halting problem by finite means, and therefore should be abandoned as problems of numerical analysis.

To obtain these results we need to agree on what it means to do something "by finite means." We use Turing's method and interpret this to mean that the "something" can be done by a certain ideal computer. That is, it can or cannot be done "by finite means" according to whether our ideal computer can or cannot do it in a finite number of steps. As has been shown by numerous developments over the past 60 years, the particular form of the ideal computer does not matter as long as the ideal computer is given basic capabilities. One might say that one ideal computer can do whatever another ideal computer can do, although the two ideal computers may do it differently.

In Turing's groundbreaking paper, the ideal computer was primitive, and the proofs of impossibility somewhat intricate. We choose an ideal computer with certain powers that an ordinary computer usually has, in order to make the various proofs of impossibility a little simpler.

In this chapter our main objective is to show that certain computation problems cannot be solved in all cases "by finite means." Any problem that *can* be solved in all cases "by finite means" is a *solvable* problem. To show that a problem is solvable is relatively easy. We need only exhibit the ideal computer solution program, which in this book we begin doing in Chapter 5. Until then we describe only the general solution procedure an ideal computer program could follow.

3.2 Turing's Resolution of the Halting Problem

The halting problem is whether there is a program for an ideal computer, which if given an arbitrary ideal computer program P and the input I accompanying it, can determine whether the program P with input I eventually terminates, that is, eventually completes the execution of its instruction list and halts.

To make this clearer we need more details about our ideal computer. It is a machine that manipulates symbol strings, and these symbol strings may be arbitrarily long. This allows the computer to compute with arbitrarily large integers of either sign, since such integers can be represented as symbol strings. The ideal computer may have arbitrarily many registers holding symbol strings, but the number of such registers N is finite. We designate these registers by the symbols $v_1, v_2, \ldots, v_N$. Initially, all registers are empty of symbol strings, except for a few of the low index registers $v_1, v_2, \ldots, v_n$, this being the "input" to the ideal computer. The "output" of the ideal computer, after it ceases computation, is the contents of another group of low index registers, $v_1, v_2, \ldots, v_m$. If P is the ideal computer program, with its registers $v_1, v_2, \ldots,$ v_n set to prescribed values $a_1, a_2, \ldots, a_n$, respectively, then $P(a_1, a_2, \ldots, a_n)$ designates its output after computation terminates, that is, the values that P leaves in $v_1, \ldots, v_m$.

Consider a present-day computer with its program occupying a specific block of B adjacent bytes of its memory. This program may be interpreted and understood if the computer's execution code is known. Similarly, there is an alphabet of symbols that can be used to specify ideal computer programs. The specific computer steps allowed are listed in Chapter 5. Thus we imagine the ideal computer is supplied its program as a string of symbols and obtains its input by having a certain number of its registers v_i set to prescribed symbol strings. Under control of its program, the ideal computer supplies the results of its computation by depositing these in a certain number of its registers before terminating program execution.

We are ready to consider the halting problem. We imagine that $P^\star$ is the ideal computer program for solving the halting problem. If P is any program that receives its input I in a single register v_1, then $P^\star$ can determine whether or not P with that input will terminate, that is, whether $P(I)$ is defined. The input to our program $P^\star$ is then a program P in v_1 and the input I to P in v_2. The program $P^\star$ can examine the symbols composing P, and so all the steps of P are available to $P^\star$. The program $P^\star$ makes its analysis of the program P with accompanying input I and supplies in v_1 a single output integer having but two values, 1 to indicate termination for P, and 0 to indicate nontermination, or endless computing by P. The $0-1$ output of $P^\star$ is then a function of P and I, which in our notation is $P^\star(P, I)$.

Thus the halting problem now is: Is there a program $P^\star$ that can do this? When a "finitely defined process" and "by finite means" are made concrete in this way, by requiring expression as a program for an ideal computer, it becomes easy to see that it is impossible to decide by finite means whether any given finite process terminates. Assuming the required program $P^\star$ is available, we obtain a contradiction. We can construct another program $P^{\neg\star}$ such that $P^{\neg\star}(I)$ is computed as follows: The program $P^\star$, stored within the program $P^{\neg\star}$, is extracted and used to compute $P^\star(P^{\neg\star}, I)$. If this result is 1, indicating termination, then $P^{\neg\star}$ enters a loop and computes forever; if this result is 0, indicating nontermination, then $P^{\neg\star}$ terminates. Thus $P^\star$ is shown to fail for at least one program P, namely $P^{\neg\star}$.

It is important to realize that there is no difficulty in deciding by finite means anything about finite processes *that terminate*, as our decision can be made by stepping through the programs defining such processes to see what they do. For instance, suppose now that $P^\star$ is a program that determines something about computations $P(I)$ *that are defined*, so that $P^\star(P, I)$ terminates after giving a 1 or a 0 as its only output, depending, say, on whether or not the v_1 output value of P is a 5. Again let us attempt to obtain a contradiction by defining $P^{\neg\star}$ to be a program such that $P^{\neg\star}(I)$ is computed by extracting the stored program $P^\star$, determining the result $P^\star(P^{\neg\star}, I)$, and then behaving in a way opposite to what $P^\star$ predicts. This time, as $P^\star$ may be assumed to do its computation by stepping through the program $P^{\neg\star}$, the program $P^{\neg\star}$ will enter an infinite loop, the computation becoming essentially a cyclic stepping of $P^\star$ through a copy of itself doing the same stepping operation. We see that $P^{\neg\star}(I)$ is undefined, and as $P^\star$ was required to decide questions only about defined computations, there is no contradiction.

Let us return to the problem of deciding whether a program halts. There is no doubt that one can construct a program $P^\star$ that can correctly determine for a huge number of cases whether or not computations $P(I)$ are defined. By analyzing at length typical programs of our ideal computer, one could no doubt make a long program P^* that would be very able. It could follow the supplied program P with the input I for an enormous number of steps to see whether P terminates, and if it did not find termination, it could then try to detect the presence in the program P of a huge variety of program loops. But the Turing proof shows that such a program P^* must always be incomplete. There is an *intrinsic* difficulty to the problem of deciding by finite means whether finitely defined processes terminate, because the assumption that such a decision procedure exists immediately leads to a contradiction.

In this chapter we show that this halting problem proof can be extended to apply to a computation involving real numbers whenever the computation requires getting endlessly more accurate approximations to the real numbers. When we obtain a particular interval approximation to a real number a, say the approximation $a(D)$, this takes only a finite number of steps, but if we keep needing better and better approximations to a, this is just like having a nonterminating program.

3.3 A Certain Computation Problem

Arithmetic operations in the field of rational numbers are relatively simple compared to the same operations in the field of real numbers. We compute $p_1/q_1 \circ p_2/q_2$ according to the rules

$$\frac{p_1}{q_1} + \frac{p_2}{q_2} = \frac{p_1q_2 + p_2q_1}{q_1q_2}$$

$$\frac{p_1}{q_1} - \frac{p_2}{q_2} = \frac{p_1q_2 - p_2q_1}{q_1q_2}$$

$$\frac{p_1}{q_1} \times \frac{p_2}{q_2} = \frac{p_1p_2}{q_1q_2}$$

$$\frac{p_1}{q_1} \div \frac{p_2}{q_2} = \begin{cases} \dfrac{p_1q_2}{q_1p_2} & \text{if } p_2 \neq 0 \\ \text{Division error} & \text{if } p_2 = 0 \end{cases}$$

There is never difficulty deciding whether two rationals are equal. We need only subtract one from the other according to the preceding subtraction rule, and if the result has a zero numerator, the two rationals are equal, otherwise they are unequal.

For two real numbers a and b, there is no simple way of determining whether or not they are equal. True, we can obtain rational approximations to them that are as accurate as we please, but consider this situation: We are trying to decide whether a equals $\frac{1}{3}$, and the approximations we obtain are:

$$
\begin{aligned}
&a(D) \\
a(1) &= 0.3^{\pm}1 \\
a(5) &= 0.33333^{\pm}1 \\
a(10) &= 0.3333333333^{\pm}1 \\
a(20) &= 0.33333333333333333333^{\pm}1
\end{aligned}
$$

As soon as we obtain a result interval that does not contain the rational $\frac{1}{3}$, perhaps the result $0.3333333331^{\pm}1$, it becomes certain that $a \neq \frac{1}{3}$. But suppose we continue to obtain results like those shown above, every approximation interval containing $\frac{1}{3}$? When do we stop computation? We must stop sometime, otherwise we would be in an unending loop if it turned out that a equaled $\frac{1}{3}$. Moreover, at whatever point we stop, perhaps for D equal to a hundred or a thousand or an arbitrarily large number, there is always the possibility that if we had continued further, finding $a(D)$ with D taken even larger, we would have found that a was unequal to $\frac{1}{3}$. Obviously then, in order to be able to determine *with certainty* whether a equals $\frac{1}{3}$, we need to examine the program $a(D)$ itself!

But how can we do this systematically for a general real number a? For all we know the number a may be computed according to the plan $\frac{1}{3} \cdot (\sin^2 \frac{\pi}{3} + \cos^2 \frac{\pi}{3})$, and so equals $\frac{1}{3}$, (or, alternately, according to the plan $\frac{1}{3} \cdot (\sin^2 \frac{\pi}{3} + \cos^2 \frac{\pi}{3}) - 10^{-1000}$, and so is unequal to $\frac{1}{3}$). To be certain of either case we would need to examine the $a(D)$ algorithm in order to recognize the identity $\cos^2\theta + \sin^2\theta = 1$ being used. There are an infinite number of other mathematical identities that also could have been employed.

The problem of deciding equality can be posed as a task for our ideal computer. Is there a program that can decide whether any two real numbers a and b are equal? Here we use the notation $a()$ to designate an ideal computer program for an approximation algorithm of type D for the real number a. It would make no difference if we specified a type K or type E algorithm instead, because from one of the three types, the other two types can be obtained constructively. Therefore the specific question now is: Is there a program $P^\star$ for our ideal computer such that $P^\star(a(), b())$ is always defined for programs $a(D)$ and $b(D)$ defining real numbers a and b, such that $P^\star(a(), b())$

equals an integer which in some agreed upon code tells us whether or not a equals b?

3.4 Deciding Whether a Number Is Zero

We may restrict ourselves to the simpler problem of deciding whether an arbitrary real number a equals zero. If there is a program $P^\star$ such that in an agreed upon code $P^\star(a())$ correctly decides whether or not a equals 0, then a program to decide whether two real numbers a and b are equal also can be constructed. This second program forms the program $c(D) = (a - b)(D)$ from its input programs $a(D)$ and $b(D)$, and then computes $P^\star(c())$ using a $P^\star$ subroutine. The value returned determines whether $a - b$ is zero, and this determines whether a equals b.

We assume now that there exists a program $P^\star$ for an ideal computer which when given the program $a()$ for an arbitrary real number a can decide whether or not a equals zero. We may suppose that $P^\star$ gives an output of 0 if a is zero, and gives an output of 1 if a is nonzero.

We define a real number z with approximation program $z(D)$ such that $P^\star(z())$ is wrong! First, we construct a general purpose program P_S, whose input is two programs P_1 and P_2, and an integer k. The program $P_S(P_1, P_2, k)$ follows the actions of $P_1(P_2)$ step by step through k steps and then supplies three output integers. The first output integer is 0 if $P_1(P_2)$ does not terminate after k steps, and is 1 if $P_1(P_2)$ terminates in k or fewer steps. If the first output integer is 1, the remaining two output integers are significant, one giving the exact number of steps N taken by $P_1(P_2)$ to termination, and the other giving the $P_1(P_2)$ output integer left in v_1. The program $z(D)$ employs P_S as a subroutine and behaves as follows: It calls $P_S(P^\star, z(), D)$, and if P_S signals nontermination of $P^\star(z())$ in D steps, or signals termination with the output 1 (meaning $z \neq 0$), then $z(D)$ gives the D decimal approximation $0.00\cdots0^{\pm}1$, that is, 0 ± 10^{-D}. If P_S signals termination with 0 output (meaning $z = 0$), then $z(D)$ makes use of N, the signaled number of steps to termination, and gives the D decimal approximation $0.00\cdots00100\cdots0^{\pm}1$ with the decimal digit 1 in position N, that is, the result $10^{-N} \pm 10^{-D}$. Thus z equals either 0 or some number 10^{-N}, depending on whether $P^\star$ claims z is nonzero or claims z is zero, respectively. Whatever result $P^\star$ determines for z, the program $P^\star$ is wrong. And should $P^\star$ never terminate, it still is wrong, because it failed to give a result for a valid approximation program $z()$, the number z being equal to 0 in this case. This shows that the problem of deciding whether a real number is zero cannot be determined in all cases by our ideal computer.

3.5 A List of Nonsolvable Problems

When it has been determined that no ideal computer program exists that can solve all cases of a certain mathematical problem, then the problem is called *nonsolvable*. Thus our first nonsolvable problem is

NONSOLVABLE PROBLEM 3.1: For any real number a, decide whether or not a equals 0.

As we have seen, this result implies that the more general problem that follows is nonsolvable:

NONSOLVABLE PROBLEM 3.2: For any two real numbers a and b, decide whether or not $a = b$.

This last nonsolvable problem implies in turn two others:

NONSOLVABLE PROBLEM 3.3: For any two real numbers a and b, decide whether or not $a > b$.

NONSOLVABLE PROBLEM 3.4: For any two real numbers a and b, decide whether or not $a \geq b$.

If the first of these last two problems were solvable, then we could use the decision program to decide whether or not $a = b$ by obtaining first a decision for $a > b$ and afterwards a decision for $b > a$. A negative decision for both cases implies $a = b$. Thus Problem 3.3 cannot be solvable, otherwise Problem 3.2 would be too. By similar reasoning, Problem 3.4 is nonsolvable, and likewise the problem that follows.

NONSOLVABLE PROBLEM 3.5: For any two real numbers a and b, choose one of the following relations as true: $a > b$, $a = b$, or $a < b$.

This last problem sometimes arises when attempting to approximate a real number a to k *correct* decimal places. The usual interpretation of a *k decimal approximation correct to the last decimal place* is that the error of the approximation does not exceed a half unit in the last decimal place. To indicate that a decimal approximant to some real number is correct to the last decimal place, it is convenient to attach the symbol $\sim$ at the end of the number as a reminder that the error bound is half a unit in the last place. Thus 54.1234$\sim$ indicates that the error bound is 0.00005. Whenever the symbol $\sim$ appears, one can replace it by $^{\pm}\frac{1}{2}$ to indicate the half unit error bound. For example, 54.1234~ is $54.1234^{\pm}\frac{1}{2}$, the decimal significance of "$\frac{1}{2}$" matching the last decimal place of the approximant.

To see how Problem 3.5 might arise, suppose we are trying to form a correct three-decimal approximation to a number a, and using $a(D)$, we obtain

$$\begin{aligned} &a(D) \\ &a(3) = 1.111^{\pm}1 \\ &a(4) = 1.1115^{\pm}1 \\ &a(5) = 1.11150^{\pm}1 \\ &a(6) = 1.111500^{\pm}1 \\ &a(7) = 1.1115000^{\pm}1 \\ &\ \ \vdots \qquad\quad \vdots \end{aligned} \tag{3.1}$$

If for higher and higher D we continue to obtain approximation intervals that contain the number 1.1115, then we are uncertain whether to use 1.111~ or 1.112~ to indicate the correct three-decimal approximant. Here we need to determine whether a is > 1.1115, is equal to 1.1115, or is < 1.1115. In the first case we can use 1.112~, in the last case we can use 1.111~, and in the middle case we can use either 1.112~ or 1.111~. A similar situation may arise when the required number of correct decimals is any value, so we have

NONSOLVABLE PROBLEM 3.6: For any real number a and positive integer k, determine a correct k decimal approximant.

3.6 Solvable Problems

In numerical analysis, it is helpful to be aware of nonsolvable problems, and to avoid trying to accomplish any process that is equivalent to attempting a nonsolvable problem. If an ideal computer cannot always solve a certain problem, then an attempt to solve the problem in full generality with merely a conventional computer is sure to have some flaw. However, often a nonsolvable problem becomes solvable if we restrict the problem in some way. Thus with the three choices of Nonsolvable Problem 3.5, it is the equality decision that is the central difficulty. If we can replace equality with an appropriate "approximate equality," the difficulty disappears. A possibility along this line is

SOLVABLE PROBLEM 3.7: For any two real numbers a and b and a positive integer k, choose one of the following as true:

$$a > b,\ a - b = 0.00\langle k \text{ zeros}\rangle 00\sim,\ \text{or } a < b.$$

Here we are using for the first time $0.00\langle k \text{ zeros}\rangle 00$ to denote the number $0.00\cdots 00$ with k zeros following the decimal point. Thus the middle choice is equivalent to

$$|a - b| \leq 0.00\langle k \text{ zeros}\rangle 005 = 0.5 \cdot 10^{-k}$$

Note that for this problem it is possible that two of the three choices are true (the middle choice and one of the others), in contrast to Problem 3.5, where only one of the three choices could be true.

To solve the preceding problem we use the decimal approximation algorithm for $a - b$ and obtain

$$\begin{aligned}&(a-b)(D)\\&(a-b)(k+1) = d_0 \,.\, d_1 d_2 \cdots d_{k+1}{}^{\pm}1\end{aligned}$$

Only if $d_0 \,.\, d_1 d_2 \cdots d_{k+1}$ equals $0.00\langle k+1 \text{ zeros}\rangle 00$, or $0.00\langle k \text{ zeros}\rangle 01$, or $-0.00\langle k \text{ zeros}\rangle 01$, does the interval $(a - b)(D)$ contain 0, and in these cases we choose the middle option. Otherwise $a - b \neq 0$, and dependent upon whether we find the approximation interval is positive or negative, we choose the first or third possibility.

The last nonsolvable problem of the previous section is especially easy to convert to a solvable form:

SOLVABLE PROBLEM 3.8: For any real number a and positive integer k, determine a to k correct decimal places or to $k + 1$ correct decimal places.

This time we obtain

$$\begin{aligned}&a(D)\\&a(k+2) = d_0 \,.\, d_1 d_2 \cdots d_{k+1} d_{k+2}{}^{\pm}1\end{aligned}$$

If $d_{k+1}d_{k+2}$ is not 50, then we can round the approximant to k decimal places to satisfy the problem. In this case the rounding introduces an error of at most 49 units in decimal place $k + 2$, and this error plus the approximant's error bound of 1 unit in decimal place $k + 2$ will not exceed 50 units in decimal place $k + 2$, or a half unit in decimal place k, as required. If $d_{k+1}d_{k+2}$ is 50, then the approximant can be rounded to $k + 1$ decimal places with no rounding error, and then the total error, being no greater than 1 unit in decimal place $k + 2$, is well under half a unit in decimal place $k + 1$. Thus for the example depicted in Eq. (3.1), where three correct decimals are desired, the $a(5)$ approximant can be rounded to give the four decimal value $1.1115\sim$, and in this way we avoid having to decide whether the number a is greater than, equal to, or less than 1.1115.

In later chapters we make use of the fact that even if a $k + 2$ decimal approximant has an error as large as 3 units in its last place, one still can obtain

k or $k+1$ correct decimal places from it. Here if $d_{k+1}d_{k+2}$ is not 48, 49, 50, 51, or 52, we round the approximant to k places, and this time the rounding error is at most 47 units in decimal place $k+2$, keeping the total error to no more than 50 units in decimal place $k+2$ or half a unit in decimal place k, as required. Otherwise we round the approximant to $k+1$ decimal places and obtain the terminal digit 5, and the rounding error now is no more than 2 units in decimal place $k+2$, keeping the total error to no more than 5 units in decimal place $k+2$, or half a unit in decimal place place $k+1$, as required.

Of course another way to convert Problem 3.5 to a solvable problem would be to restrict the real numbers a and b to rational numbers that are given as quotients of two integers. The problem would then be

SOLVABLE PROBLEM 3.9: For any two rational numbers $a = p_a/q_a$ and $b = p_b/q_b$, choose one of the following as true: $a > b, a = b$, or $a < b$.

All of the nonsolvable problems so far listed can be converted to solvable problems in this manner if the real numbers are supplied as rationals in ordered integer form. Note, however, that we obtained the contradiction to the existence of a program to solve Problem 3.1 by using a number z that is known to be rational, so with Problem 3.1, knowing that the real number a is rational without knowing its ordered integer form does not suffice to convert the problem to solvable form.

It is more of a challenge to convert a nonsolvable problem to a solvable one without restricting computation to rational arithmetic. Here is another solvable alternative to Problem 3.5:

SOLVABLE PROBLEM 3.10: Given two distinct real numbers b_1 and b_2, for any real number a choose as true one of the following inequalities: $a > b_1$, $a < b_1, a > b_2, a < b_2$.

Because we presume $b_1 \neq b_2$, we have integers K_1 and K_2 (or can obtain these) such that the intervals $b_1(K_1)$ and $b_2(K_2)$ do not intersect. We then choose an integer D such that $2 \cdot 10^{-D}$, the length of the interval $a(D)$, is less than the distance separating the interval $b_1(K_1)$ from the interval $b_2(K_2)$. The interval $a(D)$ cannot intersect both $b_1(K_1)$ and $b_2(K_2)$. If we find $a(D) \neq b_1(K_1)$, then we are able to determine whether $a > b_1$ or $a < b_1$. If we find $a(D) \neq b_2(K_2)$, then we are able to determine whether $a > b_2$ or $a < b_2$.

3.7 Key Nonsolvable Problems

As we have seen, Nonsolvable Problem 3.1 can be used to show that many other mathematical problems are nonsolvable too. To show that any of these other problems is nonsolvable we do not need the detailed proof given for Problem 3.1, but need only show that a proposed solution program for the problem leads to a solution program for Problem 3.1, an impossibility. Therefore, Problem 3.1 turns out to be a key nonsolvable problem, implying that many other problems are nonsolvable.

There are two other key nonsolvable problems we need. All three key problems concern a single real number a, and each problem may require for its resolution an unending series of values from the approximation algorithm $a(D)$. Consider first Problem 3.1, determining whether a is zero. Here the unending series of values is needed whenever a equals 0, this fact not being known to us. Obtaining $a(D) \doteq 0$ for a particular D, and being unable to decide whether a is or is not zero, our only recourse in general is to increase D and obtain a better approximation. Therefore, the difficulty of needing better and better a approximations occurs only in certain cases of a. The unending series of approximations makes the resolution of the problem like trying to decide whether an ideal computer program terminates or not.

Each of the next two sections introduces a key nonsolvable problem.

3.8 Deciding Whether a Number Is Rational or Irrational

Suppose we are given a real number a and wish to determine whether a is rational or not. Using the approximation algorithms $a(K)$ or $a(E)$, if we ever obtain an approximation interval $m \pm 0$ for a, then we know that a is rational. And if we find, by examination of the a algorithm, that the number being approximated is a known irrational number like $\sqrt{2}$ or π, then a is irrational. In general, however, without these special circumstances, we would be in a quandary as to whether a was or was not rational. In a sense, for any number a for which the fortunate circumstances just mentioned do not occur, we need the infinite decimal expansion of a to make our decision. If we could somehow view this infinite decimal expansion, then if we see that the digits repeat, the number is rational; otherwise the number is irrational. With only a finite decimal expanion for a, we cannot be certain that a cyclic repetition of digits will persist, and if there is no cyclic repetition, we cannot be certain that a cyclic repetition will not eventually appear.

For this problem we now assume there is an ideal computer program P^* that can solve this problem. Its input is again the program $a(D)$ of some real number a, and its output is a single integer giving its decision. We may assume that if a is rational, then $P^*(a())$ equals 1, and if a is irrational, then $P^*(a())$ equals 0.

Again we make use of the subroutine $P_S(P_1, P_2, k)$, which follows the actions of P_1 with input P_2 through exactly k steps, and then signals whether or not $P_1(P_2)$ terminates by this number of steps. If P_1 does terminate, then P_S also signals the exact number of steps P_1 took to terminate, and what the P_1 output was. We define a real number z with approximation program $z(D)$ that uses this subroutine and behaves as follows: It calls $P_S(P^\star, z(\,), D)$. If P_S signals $P^\star(z(\,))$ termination in exactly N steps with $P^\star$ output 1 (z is rational), then $z(D)$ returns the D decimal place number which begins with $0.00\cdots 0$ to $N-1$ decimal places, but then has in decimal places N through D the digits of $\sqrt{2}$, that is, the successive digits $14142135\ldots$, etc. Thus in this case $z(D)$ defines the number $\sqrt{2}\cdot 10^{-N}$. For all other cases $z(D)$ returns a D decimal place zero with an error bound of 1 unit in its last place. Thus, whatever decision $P^\star$ gives for z, it is wrong. If $P^\star(z(\,)) = 0$ (z is irrational), it is wrong because z equals the rational number zero. If $P^\star(z(\,)) = 1$ (z is rational), it is wrong because z is a rational multiple of $\sqrt{2}$, an irrational number. If $P^\star$ does anything else besides giving a 0 or 1 output and halting, then it has failed to give the agreed upon result for a real number z that equals the rational number zero. This shows that the problem of deciding whether a real number is rational is nonsolvable.

Accordingly we may now add to our list of nonsolvable problems the following:

NONSOLVABLE PROBLEM 3.11: For any real number a, decide whether or not a is rational.

3.9 Deciding Which of Two Real Numbers Is Larger

The problem to be considered here is whether, given two real numbers a and b, one can decide which is larger. If the two numbers turn out to be equal, it is permissible to indicate either of the numbers as the larger. Thus we either decide that $a \geq b$ or we decide that $b \geq a$. We need consider only a special case of the problem, that of determining for any number a whether a or 0 is larger, that is, either decide that $a \geq 0$ or decide that $0 \geq a$. If this simpler problem is solvable, then so is the more general one, because for two numbers a and b we can choose a as larger if $a - b \geq 0$, and we can choose b as larger if $0 \geq a - b$. If the simpler problem is nonsolvable, then the more general problem is too.

For our simpler problem, once more we assume there is an ideal computer program $P^\star$ that can solve it. Its input is the approximation program $a(D)$ of some real number a, and its output is 1 if it decides $a \geq 0$, but 0 if it decides

$0 \geq a$. If a equals 0, then $P^\star$ may give either output and be counted as giving a correct answer.

Once again we make use of the subroutine $P_S(P_1, P_2, k)$, which follows the actions of P_1 with input P_2 through exactly k steps, and then signals whether or not $P_1(P_2)$ terminates by this number of steps. If P_1 does terminate, P_S also signals the exact number of steps N that P_1 took to terminate, and what the P_1 output was. We define a real number z with the approximation program $z(D)$ that uses P_S as a subroutine and behaves as follows: First it calls $P_S(P^\star, z(\,), D)$. If P_S signals $P^\star(z(\,))$ termination in N steps with output 1 (meaning $z \geq 0$), then $z(D)$ equals the D decimal result $-0.00\cdots00100\cdots0^{\pm}1$ with the decimal digit 1 in position N, that is, $z(D)$ equals the interval $-10^{-N} \pm 10^{-D}$. If P_S signals $P^\star(z(\,))$ termination in N steps with output 0 (meaning $0 \geq z$), then $z(D)$ equals the D decimal result $0.00\cdots00100\cdots0^{\pm}1$ with the decimal digit 1 in position N, that is, $z(D)$ equals the interval $10^{-N} \pm 10^{-D}$. For all other cases $z(D)$ equals the D decimal result $0.00\cdots0^{\pm}1$, that is, the interval 0 ± 10^{-D}. Thus, whatever result $P^\star$ supplies, it is wrong. If $P^\star$ indicates $z \geq 0$, it is wrong because z equals a negative rational. If $P^\star$ indicates $0 \geq z$, it is wrong because z equals a positive rational. If $P^\star$ does anything else besides giving a 0 or 1 output and halting, then it has failed to give the agreed upon answer for a real number z that happens to equal zero. This shows that the problem of choosing the larger of two numbers is nonsolvable, and, accordingly, we have

NONSOLVABLE PROBLEM 3.12: For any real number a, decide that $a \geq 0$ or decide that $a \leq 0$.

As a consequence, we have also

NONSOLVABLE PROBLEM 3.13: For any two real numbers a and b, choose the larger number, choosing either as the larger if they are equal.

CHAPTER 4

Sequences and Functions

4.1 Sequences of Real Numbers

A sequence a_n is a concept needed in calculus, and according to the precepts of computable calculus, must be defined as a program for an ideal computer. Here the appropriate program is easily found:

DEFINITION 4.1: *A sequence a_n is given by a program $P(n)$ that for any positive integer n returns a K approximation algorithm. The real number a_n defined by the algorithm is called the n*th term *of the sequence.*

Occasionally one encounters sequences with n starting at 0, or perhaps starting at a certain positive value, say 3. Sometimes we also use a sequence a_n that is defined for only a finite number of values n. The necessary adjustments for such cases are obvious. Unless specified otherwise, a sequence a_n is understood to be defined for n equal to any positive integer.

The sequences ordinarily encountered in an elementary calculus course are sequences that would satisfy our definition. It is also clear that if we have a sequence a_n, we can construct from the a_n program other sequences, for instance, sequences such as

$$s_n = \sum_{i=1}^{n} a_i \quad \text{or} \quad p_n = \prod_{i=1}^{n} a_i$$

Our a_n program returns a K approximation algorithm for any term, so the s_n or p_n programs can use the a_n program as a subroutine, and for any input integer n obtain approximation algorithms for $a_1, a_2, \ldots, a_n$, and then combine these to form an s_n or p_n approximation algorithm by the method described in Section 2.4. In similar fashion, from two sequences a_n and b_n we can form a sequence c_n that equals $a_n + b_n$, $a_n - b_n$, or $a_n b_n$. And if for all n we have $b_n \neq 0$, then c_n also may equal a_n/b_n.

A more difficult question is whether we can have a sequence r_n with every rational number as a term. The required computable function $P(n)$ for r_n could be the following: Given the input argument n, a list L of rationals $r_1, r_2, r_3, \ldots,$ is constructed in a manner to be described, until the lengthening list has an nth rational, $r_n = p/q$, and then the program returns $(p/q)(K)$, that is, returns the approximation algorithm yielding the point interval $(p/q) \pm 0$ for every K. The list L is made as follows: r_1 is 0, and to obtain further list members, in succession, for the integer k equal to 1, 2, 3, ..., we examine the rationals $k/1, -k/1, k/2, -k/2, \ldots, k/k, -k/k$, to find the ones that are in reduced form (no positive common divisors of numerator and denominator other than 1). Whenever a reduced form rational is found, it is added to the end of the list L so far created. Every rational eventually finds its place on the list, because a rational in reduced form p/q, with q positive, is found in the k examination sublist after k reaches $\max(|p|, q)$. And this rational is never added a second time to the list L, because it is not in reduced form whenever it appears again on a k examination sublist. Thus there is a sequence r_n whose terms are distinct rational numbers, and such that every rational number is a term.

4.2 The Cantor Counting Theory

Interestingly, the beginnings of the Cantor counting theory can be presented in a computable calculus setting. A set of real numbers is *countable* or *denumerable* if there is a sequence a_n with the following properties: Every member of the set is a sequence term, with $a_n \neq a_m$ if $m \neq n$. Thus the set of positive integers obviously is denumerable, the set of all integers is denumerable, and, as just shown, the rational numbers are denumerable. If we assume that the real numbers themselves are denumerable, we encounter Cantor's famous argument that this is impossible: Let the sequence for all the reals be b_n defined by a program $P(n)$. Now let c be the decimally expressed number $0.\widehat{d_1}\widehat{d_2} \cdots \widehat{d_n} \cdots$ with $\widehat{d_n}$ chosen as follows: If the D type interval $b_n(D)$ for $D = n$ is $d_0.d_1 \ldots d_n{}^{\pm}1$, the integer $\widehat{d_n}$ equals d_n plus 5 modulo 10. The number c obviously cannot equal any term b_n, but clearly for c there is a computable, decimal approximation algorithm $c(D)$ obtainable from the b_n program $P(n)$, so the assumption that the real numbers are denumerable leads to a contradiction.

However, the set of all possible programs for our ideal computer is denumerable, that is, it is possible to computably associate ideal computer programs with the positive integers. This will become clearer in the next chapter when our ideal computer is made explicit. In Chapter 3 we saw how any ideal computer program can be described by a symbol string using some alphabet of S symbols. We can view this symbol string as an integer in the S-ary number system, and in this way associate a positive integer with the program. Conversely, any positive integer n can be converted into the S-ary system to specify a program P. This is the means whereby the positive integers can be associated with ideal computer programs. For any particular integer n, the associated program may not be an allowable program, that is, the specification may be faulty in some way, but this is of little consequence. Each allowable program P has a distinct associated S-ary integer, which we indicate with the symbol $<P>$, and all allowable programs are on the computably generated list of programs that can be constructed for the positive integers, using the S-ary numbering system.

Any real number a has an approximation program $a(K)$ that appears on the program list, so the real numbers may be considered to be included within another set which is denumerable, and this makes it difficult to continue the Cantor theory in the form Cantor gave. However, our argument does illustrate one point. From the viewpoint of conventional real analysis, computable mathematics can only cover a small subset of conventional analysis, as it misses an enormous set of conventionally defined real numbers. There must be many real numbers missing from computable analysis, because the conventionally defined real numbers are not denumerable, and cannot be included within any denumerable set. All mathematical entities defined in computable analysis can be shown by a similar argument to be but a small subset of the corresponding entities possible with conventional analysis.

On the other hand, the real numbers of conventional analysis not found within computable analysis must lack approximation algorithms, and, therefore, are unlikely to be missed in a computation setting, for instance, in numerical analysis.

4.3 Functions

Now that sequences of real numbers have been defined as certain ideal computer programs, we can consider defining a function $f(x)$. What ideal computer program should we require for $f(x)$? The concept of a function is that of an association between argument x and function value $f(x)$. An argument x is supplied, and a function value $y = f(x)$ is returned. For the argument x

we have an ideal computer program $x(K)$ from which an x approximant may be obtained, as accurate as we please. A function requires that we somehow arrive at an ideal computer program $y(K)$ giving y approximants that likewise can be made as accurate as we please. The appropriate ideal computer program for $f(x)$ is $P(x())$, similar to the program $P(n)$ needed for a sequence a_n, except that the integer argument n is replaced by the argument $x()$, that is, a program defining $x(K)$. We specify to the program P a K approximation program $x()$ defining x, and the program P returns a K approximation algorithm defining a number y.

DEFINITION 4.2: *(Markov [24].) A function $f(x)$ with domain interval I is given by a program $P(x())$, such that when the argument $x()$ is set to the K approximation algorithm of a real number in I, the program P returns the approximation algorithm $y(K)$ of a real number y, the number y called the* function value *$f(x)$. Consistency is required, in that if the numbers x and x', defined by programs $x(K)$ and $x'(K)$, are such that $x = x'$, then the corresponding programs $y(K)$ and $y'(K)$ are such that $y = y'$.*

When we evaluate some function numerically, for instance, find an approximation to $f(x) = x^2 + 5$ at $x = \pi$, this operation may be thought of as composed of two steps: obtaining from the $f(x)$ program an approximation algorithm $y(K)$ for the y function value, $\pi^2 + 5$ in our example, and then evaluating $y(K)$ for a particular K.

For sequences we devoted little space to the question of which sequences of conventional calculus are computable sequences, but for functions we need to treat more carefully the corresponding question of which functions of conventional calculus are computable. The consistency requirement is very stringent. There are many functions introduced in conventional calculus that fail to fit our definition. We begin by describing some types of functions that are computable.

There is little difficulty in constructing an appropriate ideal computer program when $f(x)$ is a polynomial $p(x)$ of the general form

$$p(x) = a_q x^q + a_{q-1} x^{q-1} + \cdots + a_2 x^2 + a_1 x + a_0, \quad a_q \neq 0 \tag{4.1}$$

The integer q, the *degree* of the polynomial, may be zero, in which case the polynomial reduces to a nonzero constant. We may assume we have available a program for the coefficients a_n, which we can use to obtain a K approximation algorithm for any term. A program for $p(x)$ uses the supplied $x(K)$ program and the approximation algorithm programs for $a_0, a_1, \ldots, a_q$, and then composes a K approximation algorithm for $y = p(x)$ according to the preceding formula, with the rational operations executed as described in Section 2.4. After the $p(x)$ program returns $y(K)$, this algorithm can be used to obtain intervals $y(1)$, $y(2)$, etc., which can be made as accurate as we please by taking K sufficiently large. If we wanted a $y(D)$ or $y(E)$ algorithm instead, this is easily obtained by the constructive conversion method described in Chapter 2.

The needed intersection property for the real number y, namely $y(K_1) = y(K_2)$, and the needed consistency requirement for a function, namely $x = x'$ implies $y = y'$, both follow from the properties of interval arithmetic.

A polynomial is defined for any real number x, that is, the domain interval I can be taken as $(-\infty, \infty)$. Consider now the program for a function of the form $f(x) = p_1(x)/p_2(x)$, where both $p_1(x)$ and $p_2(x)$ are polynomials. Such a function $f(x)$ is called a *rational* function. The procedure described for computing a polynomial can be used here too, now that the interval arithmetic division operation has been adjusted to always give a result interval. An $f(x)$ program, defined over a domain interval I that does not contain a zero (root) of the denominator polynomial, can be obtained by using the rational operations on the approximation algorithms for x and the coefficients of the two polynomials. For example, the simple function $f(x) = (x^2 + 2)/(x - 4)$ is defined in the interval $(-\infty, 4)$ or in the interval $(4, \infty)$, and the approximation algorithm for y could be composed as $((x \cdot x) + 2)/(x - 4)$, using four rational operations.

With polynomials, we use only the rational operations for $+$, $-$, and $\times$. For the rational functions, we need also the division operation. In order to employ function formulas that use the absolute value symbol, such as $|x - 4|$ or $|x/(x - 1)|$, we need an additional operation, that of taking absolute values, defined earlier in Section 2.5.

Sometimes we also want $\max(f(x), g(x))$ or $\min(f(x), g(x))$, where $f(x)$ and $g(x)$ are known computable functions, and if we define an appropriate max and min interval operation, we can use this to define the max and min operation on real numbers. However, the identities

$$\max(a, b) = \frac{a+b}{2} + \frac{|a-b|}{2} \quad \text{and} \quad \min(a, b) = \frac{a+b}{2} - \frac{|a-b|}{2}$$

show that we can express these two operations in terms of others already given.

Next consider the wider class of algebraic functions, such as the function $f(x) = \sqrt{x^2 - 1/x}$. An algebraic function $f(x)$ is any function that can be constructed combining a finite number of polynomials in x, using the four rational operations $+$, $-$, $\times$, $\div$, plus the operation of taking an nth root. An algebraic function can be generated by the procedure of combining with rational operations the K approximation algorithms for x and the coefficients of the polynomials, if we can find an appropriate way to obtain an approximation algorithm for the nth root of a number a defined by an algorithm $a(K)$. We consider this question in the next section.

4.4 The *n*th Root of a Real Number

The rational operations with real numbers were defined by specifying the K approximation algorithm of the result, the algorithm obtained from the K approximation algorithms of the operands. Similarly, the absolute value of a real number a was defined as a K approximation algorithm obtained from the K approximation algorithm for a.

For nth roots of real numbers we change our procedure. We define the nth root of a real number a by specifying a D approximation algorithm for the root, obtained from the D approximation algorithm for a. No essential difficulty is caused by this change because any algorithm type can be constructively changed to any other algorithm type. If $a(D)$ is our input algorithm, then $a(D) = d_0.d_1 \cdots d_D{}^{\pm}1$. Here d_0 is an integer that may require an arbitrary number of decimal digits to specify, while d_i for $i > 0$ is always a decimal digit with only 10 values. If the approximant is negative there is also a minus sign prefix, while if the approximant is positive or zero, there is no sign prefix. It is convenient here to assign to d_0 any sign prefix. For example, if $a(2) = -0.04$, then d_0 is to be considered to be -0.

Suppose that n is a positive *odd* integer, and we want $\sqrt[n]{a}(D) = d'_0.d'_1 \cdots d'_D{}^{\pm}1$. The nth power of $d'_0.d'_1 \cdots d'_D$ has the form $d''_0.d''_1 \cdots d''_{nD}$ with n times as many decimal digits after the decimal point. We can obtain from $a(D)$ the decimal interval $a(nD) = d_0.d_1d_2 \cdots d_{nD}{}^{\pm}1$. If we form in succession the nth powers of various D decimal place numbers, starting with $0.00\langle D \text{ zeros}\rangle 00$ and repeatedly, depending on the sign of d_0, adding 1 to or subtracting 1 from the Dth decimal digit, eventually we would be able to locate two successive D decimal place numbers, say α and β, differing only by one unit in their last decimal place, whose nth powers define an interval containing the midpoint $d_0.d_1d_2 \cdots d_{nD}$ of $a(nD)$. If the interval defined by endpoints α^n and β^n contains the entire interval $d_0.d_1d_2 \cdots d_{nD}{}^{\pm}1$, then either α or β, with the usual error bound of 1 unit in the last decimal place, can serve as the returned value for $\sqrt[n]{a}(D)$.

It may happen that $d_0.d_1d_2 \cdots d_{nD}{}^{\pm}1$ is not contained in the interval formed by α^n and β^n, but in this case we can locate three successive D decimal place numbers, say α, β and γ, with β being the middle value, such that the nth powers of α and γ define an interval containing $d_0.d_1d_2 \cdots d_{nD}{}^{\pm}1$. This time $\sqrt[n]{a}(D)$ is taken as β with the usual error bound.

Some examples may clarify the process. Suppose $n = 3$ and $D = 2$, so $nD = 6$. If $a(6) = 3.273535^{\pm}1$, we find $1.48^3 = 3.241792$ and $1.49^3 = 3.307949$, so we may take $\sqrt[3]{a}(2)$ equal either to $1.48^{\pm}1$ or $1.49^{\pm}1$. On the other hand, if $a(6) = 0.000000^{\pm}1$, then there is no choice and $\sqrt[n]{a}(D) = 0.00^{\pm}1$, because $-0.01^3 = -0.000001$, $0.00^3 = 0.000000$, and $0.01^3 = 0.000001$.

Next suppose n is a positive *even* integer. If the interval $a(nD)$ is negative, then an error must be indicated, while if the interval $a(nD)$ is positive, the previously described method can be used without essential modification. If the interval $a(nD)$ contains the zero point, it has one of three forms: $-0.00\cdots01^{\pm}1$, $0.00\cdots00^{\pm}1$, or $0.00\cdots01^{\pm}1$. We must assume that $a(D)$ is the approximation algorithm for a nonnegative real number, so can shorten the first of the three intervals to the zero point, and the second interval to $[0.00\cdots00, 0.00\cdots01]$, and then in these two cases we can take $\sqrt[n]{a}(D)$ equal to $0.00\cdots00^{\pm}1$. For the third interval we can take $\sqrt[n]{a}(D)$ equal to $0.00\cdots01^{\pm}1$. This completes the description of the process for forming $\sqrt[n]{a}(D)$.

Note that for even or odd n, the intervals $\sqrt[n]{a}(D_1)$ and $\sqrt[n]{a}(D_2)$ intersect. For if they did not, then their nth powers would not intersect, a contradiction because these nth powers contain intersecting intervals $a(nD_1)$ and $a(nD_2)$.

If the number a is defined by a K approximation algorithm, and a K approximation algorithm is required for $\sqrt[n]{a}$, this is obtained by first converting $a(K)$ to $a(D)$ in order to obtain $\sqrt[n]{a}(D)$, which is then converted to a K approximation algorithm.

4.5 An Algebra of Functions

THEOREM 4.1: If $f(x)$ and $g(x)$ are functions defined on an interval I, then the functions $|f(x)|$, $f(x)+g(x)$, $f(x)-g(x)$, $f(x)g(x)$, $\max(f(x), g(x))$, $\min(f(x), g(x))$, and $\sqrt[n]{f(x)}$ for n odd, are defined on I. If $f(x) \geq 0$ on I, then the function $\sqrt[n]{f(x)}$ for n even is defined on I. If $g(x) \neq 0$ on I, then the function $f(x)/g(x)$ is defined on I. If $h(x)$ is defined on an interval I_1, and for all x in I the function $f(x)$ has a value in I_1, then the function $h(f(x))$ is defined on I.

Let P_f, P_g, and P_h be the programs for generating $f(x)$, $g(x)$, and $h(x)$, respectively. These programs become subroutines that are used by other programs that generate the functions listed in the theorem. If $F(x)$ is $f(x) \circ g(x)$, where $\circ$ can be $+$, $-$, $\times$, or $\div$, the program P_F first obtains from P_f and P_g approximation algorithms for $f(x)$ and $g(x)$, and then returns an approximation algorithm for $f(x) \circ g(x)$. Programs for $|f(x)|$, $\sqrt[n]{f(x)}$, $\max(f(x), g(x))$, and $\min(f(x), g(x))$ can be constructed similarly.

For the last case, $G(x) = h(f(x))$, the program P_G first obtains an approximation algorithm $y(K)$ for $y = f(x)$ from a P_f subroutine, and supplies this as input to the subroutine P_h, its output becoming the output of P_G.

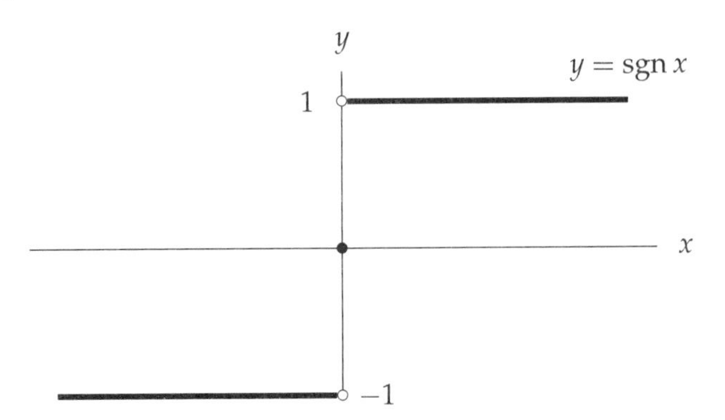

Figure 4.1. The conventional function sgn x which is 0 at $x = 0$.

4.6 The Function sgn(x)

The sign function (see Fig. 4.1), usually given the abbreviation sgn x, can be defined for rational x by the equation

$$\operatorname{sgn} x = \begin{cases} +1 & \text{if } x > 0 \\ 0 & \text{if } x = 0 \\ -1 & \text{if } x < 0 \end{cases}$$

If we try to extend this function's domain into the larger field of real numbers, we encounter difficulty. Indeed we see that if we were successful in defining sgn x for all real x, we could decide whether or not x equaled zero, contradicting Nonsolvable Problem 3.1. That is, if $\operatorname{sgn}(x) = y$, after converting $y(K)$ to $y(D)$ and setting $D = 1$, if we find the interval $y(1)$ contains the zero point, it is certain that $x = 0$; otherwise, it is certain that $x \neq 0$. Thus, it is clear that sgn x cannot be realized as a computable function. What can be realized is a sgn x function defined in $(-\infty, 0)$ and in $(0, \infty)$, but undefined for x equal to 0 (see Fig. 4.2).

For a proposed function such as sgn x, we can consider how far we can go toward its complete realization. An appropriate sgn x output interval value for an arbitrary x input interval $m \pm e$ is

$$\operatorname{sgn}(m \pm e) = \begin{cases} +1 \pm 0 & \text{if } m \pm e > 0 \\ 0 \pm 0 & \text{if } m \pm e \text{ is } 0 \pm 0 \\ 0 \pm 1 & \text{if } m \pm e \doteq 0 \\ -1 \pm 0 & \text{if } m \pm e < 0 \end{cases} \tag{4.2}$$

Using this interval relation, it is true that we can create an ideal computer program that could deliver the correct value for some x arguments equal to 0, but not all such arguments. Of course one could attempt to extend the sgn x interval relation given in the preceding by including an internal examination

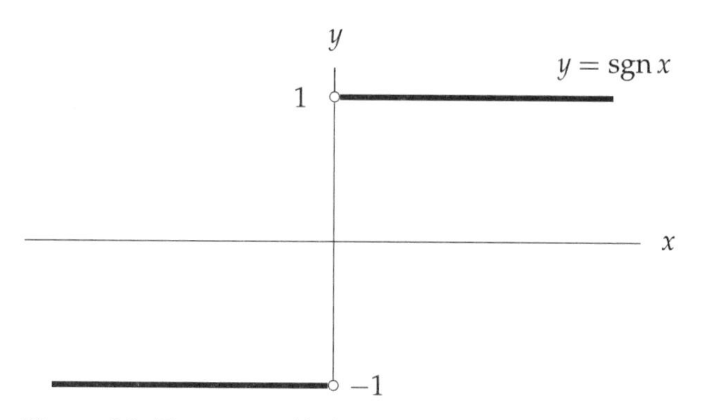

Figure 4.2. The computable function sgn x which equals $x/|x|$.

of the $x(K)$ approximation program, attempting to detect cases where the number x is zero. No doubt a sgn x program more able to detect zero arguments would result, but Nonsolvable Problem 3.1 indicates that no matter how complicated or clever the procedure for detecting zero arguments is, the function in some cases of zero arguments assigns a wrong value. In computable calculus the function sgn x must be undefined for $x = 0$ in order to avoid this difficulty for the argument $x = 0$. At all other x arguments, a value for sgn x is obtainable, because we can take sgn x as equal to $x/|x|$ if we give up trying to obtain a value for the function at $x = 0$.

For a more extreme case of this difficulty realizing discontinuous functions, consider the function $r(x)$ given here and often introduced in calculus texts as an example of a function everywhere discontinuous:

$$r(x) = \begin{cases} 1 & \text{if } x \text{ is rational} \\ 0 & \text{if } x \text{ is irrational} \end{cases}$$

Here we could begin an attempt to realize $r(x)$ by using this interval relation derived from the definition of $r(x)$:

$$r(m \pm e) = \begin{cases} 1 \pm 0 & \text{if } m \pm e \text{ is } m \pm 0 \\ 0 \pm 1 & \text{otherwise} \end{cases}$$

Further development of this realization attempt cannot completely succeed in defining $r(x)$, otherwise we contradict Nonsolvable Problem 3.11.

In Chapter 6, it is shown that any computable function $f(x)$ must be continuous at any argument in its domain. A discontinuous function of conventional calculus, which has just one or a few points of discontinuity, can still be used as a function in computable calculus, but only after the function is changed to be *undefined* at every point where it was required to be discontinuous. Thus, suppose the function $h(x)$ is defined to equal $2x$ for x in the interval $[0, \infty)$ and to equal -4 for x in the interval $(-\infty, 0)$. This function is discontinuous at $x = 0$. If we wanted to use $h(x)$ in computable calculus, we would need to change the first part of the $h(x)$ definition to make $h(x)$ equal $2x$ in the interval $(0, \infty)$. Now $h(x)$ is undefined at $x = 0$.

To gain understanding of the difference between computable functions and conventional functions, consider the following situation. Suppose it is known that the computable function $f(x)$ maps every rational number in the interval $m \pm e$ into the interval $m_1 \pm e_1$. Does this imply that every real number in $m \pm e$ is likewise mapped into $m_1 \pm e_1$? A conventional function can be any mentally conceivable function, so it would be possible for a conventional function $f(x)$ to have many real arguments in $m \pm e$ mapped outside $m_1 \pm e_1$.

However, this is not the case for a computable function $f(x)$. Assume that $z(K)$ defines some real number in $m \pm e$ that $f(x)$ maps into $y(K)$ outside $m_1 \pm e_1$. We show that this assumption is wrong because it implies a method of deciding whether any approximation algorithm $a(K)$ is equal to zero, thereby contradicting Nonsolvable Problem 1.1. If we compute in succession $y(1)$, $y(2)$, $y(3)$, ..., eventually we obtain an interval $y(K')$ outside $m_1 \pm e_1$. Let δ be the rational distance separating $y(K')$ and $m_1 \pm e_1$. If $a(K)$ is an arbitrary approximation algorithm, we can construct the approximation algorithm $z_a(K)$ which for any K first tests whether the intervals $a(1)$, $a(2)$, $a(3)$, ..., $a(K)$ all contain the zero point. If so, then $z_a(K)$ equals $z(K)$. Otherwise, if $a(K_0)$ is the first interval found to be positive or negative, then $z_a(K)$ equals a fixed rational chosen inside $m \pm e$ and inside the mutually intersecting intervals $z(1)$, $z(2)$, ..., $z(K_0)$. Thus the number z_a equals z if a is zero, and equals a rational inside $m \pm e$ if a is nonzero. If the algorithm returned by $f(x)$ for the input $z_a(K)$ is $y_a(K)$, we can find a $y_a(K)$ interval that is less than δ in length. If this interval intersects $m_1 \pm e_1$, the number z_a must be rational and so a is nonzero. Otherwise, the number z_a must equal z, so a is zero.

4.7 Functions Defined on Intervals

The interval I on which a function $f(x)$ is defined may be a finite interval of any of the four forms (a, b), $(a, b]$, $[a, b)$, $[a, b]$, or may be an infinite interval of the five forms $(-\infty, \infty)$, $(-\infty, b)$, $(-\infty, b]$, (a, ∞), or $[a, \infty)$. An interval with finite endpoints whose form employs brackets only or parentheses only is called *closed* or *open*, respectively. If the interval I is of the forms $[a, b)$, $[a, b]$, or $[a, \infty)$, then I is said to be *closed at its left endpoint*. Similarly I is *open at its left endpoint* if I is of the form $(a, b]$, (a, b), or (a, ∞). Analogous interpretations apply to an interval open or closed at its right endpoint.

Our definition of a function $f(x)$ defined on an interval I makes no mention of what happens if the x approximation algorithm supplied to $f(x)$ is for a real number that is outside I. In this out-of-domain case, is it possible to require the $f(x)$ program to signal an error, or failing that, is it possible to require the y approximation algorithm returned by $f(x)$ to signal an error as soon as it is used to obtain y intervals? This question may arise when considering how to realize a particular function $f(x)$ on an actual computer.

Consider the function $\sqrt{x}$ with the domain $[0, \infty)$, and suppose this function has the form where it accepts D algorithms as inputs and returns D algorithms as output. We also suppose that the returned $y(D)$ algorithm is computed according to the plan for obtaining nth roots given in Section 4.4. Suppose x is negative but close to zero. For instance, if x is -10^{-100}, it is possible $x(D)$ equals $0.00\langle D \text{ zeros}\rangle 00^{\pm}1$ for $D < 100$, and becomes $-0.00\langle 99 \text{ zeros}\rangle 00100 \cdots 00^{\pm}1$ for $D \geq 100$. The $\sqrt{x}(D)$ algorithm returns an interval containing the zero point when the parameter $D \leq 50$, and only for larger D when a negative interval for $x(2D)$ is obtained, can the algorithm recognize the invalid x argument and signal an error. For this method of realizing the function $\sqrt{x}$, we see that the returned $y(D)$ approximation algorithm does not always expose an out-of-domain x argument for all values of the parameter D.

A similar difficulty will occur with any function $f(x)$ with a domain interval I closed at the left endpoint a. That is, the function program itself cannot always detect that a supplied x argument is $< a$, nor can the program always return a $y(D)$ algorithm revealing the error for every D. For if we suppose that the $f(x)$ program is in either way able to identify x arguments outside $[a, \infty)$ and signal an error, we now have a method of deciding whether or not $x < a$, and this contradicts Nonsolvable Problem 3.3. A similar difficulty occurs for a function $f(x)$ with a domain interval I closed at its right endpoint b.

This quirk of sometimes obtaining an $f(x)$ value when an error should be signaled need not occur if the domain interval I is open at its left endpoint a, but another difficulty occurs instead. For instance, consider the function $1/\sqrt{x}$ with the domain $(0, \infty)$, and suppose this function returns a $y(D)$ approximation algorithm computed by the operation of taking an nth root of $x(D)$, described in Section 4.4, converting this to the K form, performing the division operation as described in Section 2.4, and then converting the quotient result back to the D form. For x negative, the returned $y(D)$ algorithm cannot be defined for any D, because the $\sqrt{x}(D)$ algorithm never gives a positive interval needed to obtain such a finite interval result. For every argument D, the $y(D)$ algorithm, perhaps after prolonged computation, eventually detects a negative $x(D)$ interval and signals an error.

But what if x is zero? This fact may perhaps be uncovered by an $x(K)$ computation that yields the point interval 0 ± 0, but according to Nonsolvable Problem 3.1, there is no general way of determining whether $x = 0$. Thus there are zero arguments x such that $(1/\sqrt{x})(D)$ for any D can never terminate in the attempt to obtain a decimal interval, which means the returned $y(D)$

approximation algorithm is undefined for every D. In general, when the $f(x)$ domain interval I is open at a, an error can always be signaled if $x < a$, but there is no way to avoid having the returned y approximation algorithm undefined for at least some x arguments that equal a. There is a similar difficulty if the domain I is open at its right endpoint.

On an ordinary computer, it is not practical to imitate an ideal computer for a function defined on an interval open at either endpoint, endlessly increasing the precision of computation for an erroneously supplied x argument happening to equal the endpoint. Doing this would cause programs to cycle endlessly for such erroneous x arguments. Instead one can terminate computation and signal an "ambiguous error," that is, indicate by an appropriate error message that a computed x interval intersects a computed a interval. However, this has the unavoidable drawback that sometimes an ambiguous error message is returned for an x argument inside the domain I.

4.8 Semifunctions

Functions have the consistency requirement that if x equals x', then the corresponding function values y and y' must also be equal. For many computations, we want an approximation algorithm $y(K)$ for every approximation algorithm $x(K)$ defining a number in some interval I, but are unconcerned whether consistency is achieved. In such cases it is more convenient to employ *semifunctions*:

DEFINITION 4.3: *A semifunction $f(x())$ with domain interval I is given by a program $P(x())$, such that when the argument $x()$ is fixed at the approximation algorithm $x(K)$ for a real number in I, the program P returns a real number approximation algorithm $y(K)$, called the* semifunction value $f(x())$.

A function is a semifunction, and a semifunction may or may not be a function, depending on whether it is consistent or not. Rather than use the unappealing semifunction notation $f(x())$ of the definition, in this book we use ordinary $f(x)$ notation for a semifunction, the word "semifunction" always indicating the lack of consistency for $f(x)$. For instance, we have a semifunction and not a function if we require $f(x)$ to return 0 if the $x(E_0)$ approximant is 0, to return 1 if the $x(E_0)$ approximant is positive, and to return -1 if the $x(E_0)$ approximant is negative. Here E_0 is some arbitrary small number, perhaps 10^{-5}, limiting the magnitude of the x approximant's error bound. This semifunction might serve to approximate the conventional function $\mathrm{sgn}(x)$ of Fig. 4.1.

As another example, suppose we try to approximate the greatest integer function of conventional calculus, defined for any x to be equal to the largest integer $\leq x$. We can construct a program that returns the largest integer $\leq$ the $x(E_0)$ approximant, where again E_0 is assigned some specific value, say 10^{-5}. In this case we have constructed a semifunction and not a function. Here we cannot expect consistency, for if the number x is close to an integer, different approximation algorithms $x(E)$ for the number could easily yield different integer values. As will become evident in Chapter 6, the only integer-valued functions defined on $(-\infty, \infty)$ are constant functions, always yielding the same integer value.

4.9 Other Calculus Concepts

With sequences and functions now defined, we can consider other concepts needed in calculus. One often encounters sequences $a_{n,m}$ with a pair of indices n, m, and the appropriate definition for this case is:

DEFINITION 4.4: *A double index sequence $a_{m,n}$ is given by a program $P(m, n)$ that for any positive integers m and n returns a K approximation algorithm for a real number.*

The concepts of function and sequence are sometimes combined. For example, the sequence of functions $f_n(x)$ may be defined by an equation such as $f_n(x) = \sum_{i=1}^{n} x^n$. An appropriate definition for this concept is:

DEFINITION 4.5: *A sequence of functions $f_n(x)$, defined on an interval I, is given by a program $P(n)$ that for any positive integer n returns another program defining a function on I.*

In calculus, functions of more than one variable are used often. The $f(x)$ definition is easily extended to two variable functions $f(x, y)$ if a domain interval I_1 for x is known and a domain interval I_2 for y is known, for instance, if the Cartesian xy domain is a rectangle. However, domains for functions of two variables are often more complicated. For instance, the function $\sqrt{1 - x^2 - y^2}$ is defined in a unit circle with center at the origin, and the function $(x^2 + y^2)/(x - y)$ is defined in the entire Cartesian plane except on the line $y = x$. Therefore, in our definition of $f(x, y)$ we need to allow more complicated domains than can be obtained by restricting each variable to an interval.

For functions $f(x_1, x_2 \ldots, x_n)$ of n variables, we can obtain a more general domain $I^{(n)}$ by allowing the intervals for the variables to be bounded in a more general fashion. The first step is to arrange the variables in a particular order: $x_{k_1}, x_{k_2}, \ldots, x_{k_n}$. The first variable x_{k_1} in the specification list is associated with an ordinary interval $I^{(1)}$ that uses bracket-parenthesis notation with two endpoint values a_1 and b_1. Here a_1 and b_1 can both be numbers, in which case $a_1 < b_1$, but $a_1 = -\infty$ and $b_1 = \infty$ are allowed too. A left parenthesis must appear with $-\infty$ and a right parenthesis with ∞. The *interior* of $I^{(1)}$ is the interval obtained when brackets are converted to parentheses.

The second variable x_{k_2} has an interval that varies with x_{k_1}, that is, the x_{k_2} interval uses bracket-parenthesis notation with endpoint *functions* $a_2(x_{k_1})$ and $b_2(x_{k_1})$. Both of these functions are defined in $I^{(1)}$, with the inequality $a_2(x_{k_1}) \leq b_2(x_{k_1})$ being true for any x_{k_1} in $I^{(1)}$, and the sharper inequality $a_2(x_{k_1}) < b_2(x_{k_1})$ being true in the interior of $I^{(1)}$. The function $a_2(x_{k_1})$ is allowed to equal $-\infty$ with a left parenthesis assigned, and $b_2(x_{k_1})$ is allowed to equal ∞ with a right parenthesis assigned, with the obvious interpretations for these choices. The intervals for the variables x_{k_1} and x_{k_2} now define a certain region $I^{(2)}$ in the $x_{k_1}x_{k_2}$ plane. The *interior* of $I^{(2)}$ is the region obtained when all brackets are converted to parentheses. (As many as 4 brackets may need to be changed.)

Similarly, the third variable x_{k_3} has an interval that varies with both x_{k_1} and x_{k_2}. This interval uses bracket-parenthesis notation with the endpoint functions $a_3(x_{k_1}, x_{k_2})$ and $b_3(x_{k_1}, x_{k_2})$. Both functions are defined in $I^{(2)}$ with the inequality $a_3(x_{k_1}, x_{k_2}) \leq a_3(x_{k_1}, x_{k_2})$ being true there, but with the sharper inequality $a_3(x_{k_1}, x_{k_2}) < a_3(x_{k_1}, x_{k_2})$ true in the interior of $I^{(2)}$. This defines a region $I^{(3)}$ in $x_{k_1}x_{k_2}x_{k_3}$ space. The *interior* of $I^{(3)}$ is the region obtained when all brackets used are converted to parentheses. (As many as 6 brackets may need to be changed.)

The plan now described for the first, second and third variable is generalized to apply to all the variables in the specification list, and the final region is $I^{(n)}$. A point $(x_1, \ldots, x_n)$ in $I^{(n)}$ is an *interior point* if it also belongs to the region obtained by converting all $I^{(n)}$ brackets into parentheses, and the point is a *boundary point* if this is not the case.

As two examples, the region $I^{(2)}$ defined by $x\colon [-1, 1]$, $y\colon [-\sqrt{1-x^2}, \sqrt{1-x^2}]$ defines the unit circle and includes the boundary of this circle, while the region $x\colon (-1, 1)$, $y\colon (-\sqrt{1-x^2}, \sqrt{1-x^2})$, $z\colon (-\sqrt{1-x^2-y^2}, \sqrt{1-x^2-y^2})$ defines the interior of the unit sphere. Similarly, the domain of the function $(x^2+y^2)/(x-y)$ consists of the unbounded region $y\colon (-\infty, \infty)$, $x\colon (-\infty, y)$ and the unbounded region $y\colon (-\infty, \infty)$, $x\colon (y, \infty)$.

This region notation allows us to consider functions with more general domains than those with each variable varying in an ordinary interval. That type of domain is still a possibility in this new notation, and can be specified with endpoint functions that are constants. When in the future we call a bounded region $I^{(n)}$ open, this is to be interpreted to mean that only parentheses are used to specify the region. Similarly, a closed bounded region $I^{(n)}$ is one where only brackets are used in its specification.

A function of any number of variables may now be defined as follows:

DEFINITION 4.6: *A function $f(x_1, \ldots, x_n)$ defined in a region $I^{(n)}$ is given by a program $P(x_1(), \ldots, x_n())$, such that when $x_1(), \ldots, x_n()$ are set to approximation algorithms defining a point $(x_1, \ldots, x_n)$ in $I^{(n)}$, the program returns an approximation algorithm $y(K)$ for a real number y, called the function value $f(x_1, \ldots, x_n)$. Consistency in each variable x_i is required, in that if $x_i()$ and $x_i'()$ are such that $x_i = x_i'$, the corresponding algorithms $y(K)$ and $y'(K)$ are such that $y = y'$.*

The region $I^{(n)}$ for $f(x_1, \ldots, x_n)$ requires for its specification certain functions of $n-1$ arguments, so the forementioned definition is recursively dependent on a preceding definition of $f(x_1, \ldots, x_{n-1})$ over a region $I^{(n-1)}$.

CHAPTER 5

The Ideal Computer

5.1 The Goal of This Chapter

In the last century mathematicians such as Kurt Gödel, Alan Turing, Emil Post, and Alonzo Church, concerned with questions of logic and the foundations of mathematics, grappled with the question of what could be achieved by finite means. Gödel began the process of obtaining negative results by showing that the foundations of mathematics will always be incomplete [17]. The ensuing further application of constructivity arguments to areas of logic settled whether there can be a mechanical method for deciding mathematical questions, a dream that had existed since the time of Leibniz. The field of mathematical logic changed in a fundamental way, becoming a stronger and more useful discipline in the process.

Numerical analysis is another field of mathematics where it makes sense to consider problems from the point of view of what can be achieved by finite means. Three fundamental nonsolvable problems were listed in Chapter 3. If we find that the problem we are trying to solve numerically is equivalent to tackling one of these fundamental nonsolvable problems, then the problem should be reexamined to see whether it can be altered in a reasonable way to become solvable. In this way we avoid computational difficulties that can never be entirely overcome.

In Chapter 3 we did not give complete proofs of nonsolvability, but only sketched the proofs. In this chapter we single out one of the three fundamental nonsolvable problems, the problem of determining whether a real number is zero, and supply the missing details. Along the way we can clarify the concepts introduced in earlier chapters.

5.2 The Various Methods of Proof

Proving that there is no general constructive method of determining whether a real number equals zero depends on a reasonable definition of what constitutes a "procedure by finite means." In the early 1930s, as mathematicians attempted to understand the concept of "finite process" or "constructive process," the realization arose that since ordinary integers can be used as the basis of any computation, one needed only to clarify what a finite process was when that process involved just the manipulation of integers. Eventually the concept of *recursive functions* was arrived at. Any process involving nonnegative integers was constructive if the process could be expressed as a recursive function computation. The rules of computation one can use are explicitly laid out in the recursive function formulation of constructivity. A process is constructive if and only if it can be expressed using these rules.

The logician Alonzo Church used an alternate system of defining a constructive process that became known as *lambda processes*. Again, the rules here are explicitly laid out and a procedure is constructive if and only if it can be expressed as a lambda process. The notation of lambda processes is still used today in certain areas of computer science.

When Turing's fundamental paper appeared, his method of defining a constructive process became widely imitated, because Turing's definition is simple and intuitively appealing. Turing defined a process to be constructive if and only if the process could be executed by one of his primitive machines, later known as Turing machines. A Turing machine executes a procedure by creating an appropriate symbol string on an arbitrarily long tape, the tape used for both computation and input and output. The tape may be imagined held by a tape-reader mechanism that accesses exactly one symbol on the tape. The machine has a finite number of internal states, and when the machine is in any of these states, it responds to reading any of a fixed alphabet of symbols on its tape in the following manner: The machine may replace the symbol currently read with another symbol, or it may leave the symbol in place; next, the machine may move the tape one symbol forward, one symbol backward, or keep the tape unchanged in position; and last, the machine, as a consequence of both its internal state and the symbol read, enters another designated internal state, whereupon the whole cycle, constituting a Turing machine "step," is repeated.

Eventually the concepts of recursive functions, lambda processes, and Turing machines were shown to be equivalent, that is, a lambda process or a recursive function computation can be done in symbol string form by a Turing machine, and any Turing machine computation can be converted into a recursive function computation or into a lambda process.

The Turing machine may be viewed as an ideal computer that is ideal in two respects. A Turing machine has a tape of undefined length on which an arbitrarily long string of symbols may be written, the symbols chosen from some arbitrary finite alphabet. The tape is eraseable, because after a symbol is written, the symbol later can be deleted and replaced by a different symbol. The memory obtained by this tape is "effectively infinite" but not infinite, since an infinite memory would violate the notion of "finite process." For any computation a Turing machine does, the memory it uses is finite, but large enough for the computation. If a finite memory of a specific size were required for an ideal computer, that size, however large it was, inevitably would become a trivial complicating factor in any consideration.

The second sense in which a Turing machine is an ideal computer is that the number of steps taken to complete a computation is never a concern, as long as the number of steps is finite. Let us imagine that a Turing machine does every computation in one minute of human time, and also that all steps are done equally quickly. Then for any computation a Turing machine does, the time needed to execute one step may become very small, but this time is never zero. Instantaneous execution of programs again would violate the notion of "finite process." The Turing machine concept allows us to distinguish clearly between a process taking an infinite number of steps and a process taking a finite number of steps.

In numerical analysis, by focusing attention on whether a process takes a finite number or an infinite number of steps, we greatly improve the numerical answers we provide. We can supply accurate numerical answers routinely instead of occasionally. By avoiding any process taking an infinite number of ideal computer steps, and using just those processes taking "only" a finite number of steps, we avoid numerous computational pitfalls. Of course some of these finite step processes are beyond achieving with the computers that are actually available to us, but then we have always been limited, and always will be limited, by our computing resources.

The ideal computer becomes the arbiter of the question of whether a certain task can be done by finite means. This becomes purely a matter of deciding whether the ideal computer can be programmed to accomplish the task in a finite number of steps.

5.3 Definition of the Ideal Computer

There is considerable freedom in the specification of the ideal computer. The first ideal computer, the primitive Turing machine, was proposed by Turing years before there were any working computers that could execute stored programs. Turing's concept is so simple that it is easy to imagine how his machines might actually be constructed, and the machines, even though simple, could do the most intricate calculations, as Turing showed in great detail in his paper [36].

However, a primitive computer must do many of its computations in roundabout ways, and developing computable calculus with a primitive machine would be tedious, forcing a reader to go through the tortures that early computer users endured getting their machines to do useful work, before programming languages made the job easier. (Turing himself made certain errors in describing his machines which he corrected in an addendum [37].) It is more convenient to use an ideal computer that is similar to our everyday computers, so that our experience programming these machines can help in understanding the ideal computer and its powers. Of course there is a Turing machine that does all the things any ideal computer does, but by beginning with a more modern machine, a simple and possibly constructible machine, we save much tedious analysis.

Imagining an ideal computer that resembles an actual computer is convenient, and easily understood these days, now that computers are everywhere. Another advantage of this approach is that it becomes easy to simulate the ideal computer on an actual computer, and in this way make the ideal computer concept more vivid. One may take some task, write an ideal computer program to execute that task, and then actually test the program to see if it works as it is supposed to. This book's CD contains software that makes such projects possible.

Our ideal computer can call subroutines. Here one might consider the parallel history of our commercial computers. The very earliest computers, the one-of-a-kind variety, were inconvenient to program. For example, plugboards were used to control the ENIAC computer of the 1940s. By the early 1950s it became routine for a commercial computer to follow a stored program in memory. Somewhat later, the convenient "subroutine call" and "subroutine return" computer instructions became commonplace. By giving the ideal computer the power to call subroutines, we make use of this hard-won experience in constructing useful computers, and our ideal computer programs become simpler.

The effectively infinite memory of a Turing machine is the tape of undefined length on which the machine writes and reads its symbols. In Turing's day, paper tape readers were common, and were used to transmit and record stock market transactions. The effectively infinite memory of our ideal computer is obtained in two ways. A program may use any finite number of "variables," where a variable holds a symbol string of arbitrary length. Each variable in memory storage capacity then is equivalent to a Turing tape. Our ideal computer stores programs in another memory that resembles the disk memory of modern computers. The ideal computer has an instruction step that stores a variable's symbol string contents as a file at the file address specified by another variable's contents. The computer has a second instruction step that retrieves the file, reconverting it to a variable's symbol string contents. Such a convenient disklike memory would hardly have been acceptable at the time Turing wrote his fundamental paper, since a programmable computer was then unknown. A Turing machine can simulate such a memory by using its effectively infinite tape appropriately, but this is not easy to show, and by postulating an ideal disk memory for our ideal computer, we eliminate a great amount of tedium in getting our ideal computer to perform useful tasks.

In the next section, we list the 11 computation steps our ideal computer performs and the symbols it uses. In the succeeding sections we show how the computer can be programmed to do many of the things that were called constructively possible in Chapters 2, 3, and 4. The last two sections of this chapter give the details of the nonsolvability proof for the problem of determining whether a real number is zero. This proof can be easily adjusted to serve as a proof for either of the two other key nonsolvable problems.

5.4 The Ideal Computer Steps

Our ideal computer manipulates an arbitrary number of symbol strings, and these symbol strings may be of arbitrary length. By its power to manipulate strings, the ideal computer can also do signed integer computations with integers of any size. The different symbol strings it uses are designated by the symbols $v_1, v_2, \ldots, v_i, \ldots, v_N$, each element v_i being called a "variable," and the index i is allowed to be arbitrarily large, so there is no limit on the number of variables that can be used.

Consider the Chapter 3 notation $P(a_1, \ldots, a_n)$, used to designate the output of program P when its input is $a_1, \ldots, a_n$. As an ideal computer begins operations, the variables $v_1, \ldots, v_n$ contain the program inputs $a_1, \ldots, a_n$, and all other variables are *empty*. That is, if the program P uses v_i and $i > n$, then v_i initially contains no symbols at all.

The ideal computer does its work by executing 11 different step types, now to be defined, and a list of the steps to be executed constitutes a program for the computer. The first step type allows the ideal computer to assign any variable a string, the string either being copied from another variable or explicitly designated within quotes.

$$\text{Step type 1: } v_i \Leftarrow v_j \quad \text{or} \quad v_i \Leftarrow \text{"symbol\_string"}$$

Here symbol_string is made up of the various symbols the ideal computer manipulates (listed later). For this step and for the steps described later that change v_i, any string that v_i has prior to the step is lost after the step is executed.

The symbol strings of any two variables may be copied and then concatenated and placed in any variable.

$$\text{Step type 2: } v_i \Leftarrow v_j + v_k$$

Thus if v_1 contains 12 and v_2 contains 34, the step $v_3 \Leftarrow v_1 + v_2$ leaves 1234 in v_3 without changing v_1 or v_2.

A variable may have either its leading or its trailing symbol stripped to become the string contents of another variable.

$$\text{Step type 3: } v_i \Leftarrow [v_j \quad \text{(leading } v_j \text{ symbol removed from } v_j\text{)}$$
$$\text{Step type 4: } v_i \Leftarrow \ v_j] \quad \text{(trailing } v_j \text{ symbol removed from } v_j\text{)}$$

After one of these two step types is executed, variable v_i contains a string consisting of a single symbol, and variable v_j has its string shortened by one symbol. When the last symbol is stripped from variable v_j, then variable v_j becomes empty, and if variable v_j is already empty prior to the step, then variable v_i also becomes empty. For instance, if v_3 contains 1234, the step $v_4 \Leftarrow v_3]$ leaves 4 in v_4 and 123 in v_3. The subsequent step $v_5 \Leftarrow [v_4$ leaves 4 in v_5 and clears v_4. The subsequent step $v_6 \Leftarrow [v_4$ clears v_6.

The ideal computer executes its steps in sequential order, except for the next three step types, which allow for a variation in this order. Two of these step types allow the ideal computer to test whether a variable string is or is not identical to a test string. The test string is either the string of a designated variable or is given within quotes.

$$\text{Step type 5: jump } j \text{ if } v_i = v_k \quad \text{or} \quad \text{jump } j \text{ if } v_i = \text{"symbol\_string"}$$
$$\text{Step type 6: jump } j \text{ if } v_i \neq v_k \quad \text{or} \quad \text{jump } j \text{ if } v_i \neq \text{"symbol\_string"}$$

If the condition being tested for occurs, then the next instruction to be executed is j steps away, where j is allowed to be any nonzero integer, positive for a forward jump, and negative for a backward jump. Sometimes it is convenient to test whether variable v_i is empty, and this can be done either with the empty quote ("") or by arranging to have the comparison variable v_k empty. If the condition being tested for does not occur, then, as usual, the next step executed is the succeeding step. The last instruction step of this group executes the jump without making any test.

Step type 7: *jump j*

The size of j in each of these three steps is restricted to values that lead to existing steps within the program. For instance, if a jump j step is the Nth step of a program, then there must be an $(N + j)$th step in the program.

The next two step types are related.

Step type 8: call program

Step type 9: return

Each program is assigned some identifying symbols that become its "name" to be used in call steps. Any program can be called by name to become a subroutine of another program. After the called program has completed its computation, it executes a return step and returns control to the calling program. Allowing a call step and a return step makes it easy to avoid dealing with a particular computation issue twice. We master dealing with the issue by creating a certain program, and then when the issue arises again later, we can deal with it by calling this program by its name.

For any program P, the number n of input variables it obtains is specified, and also the number m of output variables it supplies when computation terminates. When P is called by some other program, variables $v_1, \ldots, v_n$ of P get filled with strings copied from variables $v_1, \ldots, v_n$ of the calling program. The variables of a calling program are inaccessible to a called program, which uses its own independent set of variables. As P, the called program, executes its steps and changes its own variables, the variables of the calling program remain unchanged. Finally, when P returns control by executing a return step, its variables $v_1, \ldots, v_m$ are copied into variables $v_1, \ldots, v_m$ of the calling program, then the contents of all variables of P are lost as control is returned to the calling program, this program resuming execution at the step just after the call step. A called subroutine may call its own set of subroutines, the whole subroutine call mechanism being recursive.

The return step is used by *every* program to cease computation, not just programs which get called. When previously we designated some computation result as $P(a_1, \ldots, a_n)$, then the program P, the "main" program, also ceases computation with a return step, and when it does, its computation result, contained in its m variables $v_1, \ldots, v_m$, becomes available for our inspection.

Our ideal computer has two last step types, for "input" and "output" communication with an ideal disk memory. For the input step, the disk file with the v_1 symbol string name is copied into variable v_2, the previous contents of this variable being lost. For the output step, the disk file with the v_1 symbol string name is cleared of any previous contents, and then the v_2 symbol string is copied into this file.

We list below the ideal computer's 11 step types as they are written for human understanding, and as they would appear in a program for the computer. A program step in either representation has a terminal semicolon to separate two successive steps.

Type	Step Representation	Ideal Computer Code
1	$v_i \Leftarrow \begin{cases} v_j; \\ \texttt{"symbol\_string"}; \end{cases}$	$1\vert i\vert j;$ or $1\vert i\,\texttt{"symbol\_string"};$
2	$v_i \Leftarrow v_j + v_k;$	$2\vert i\vert j\vert k;$
3	$v_i \Leftarrow [v_j\ ;$	$3\vert i\vert j;$
4	$v_i \Leftarrow\ v_j];$	$4\vert i\vert j;$
5	jump j if $v_i = \begin{cases} v_k; \\ \texttt{"symbol\_string"}; \end{cases}$	$5/j\vert i\vert k;$ or $5/j\vert i\,\texttt{"symbol\_string"};$
6	jump j if $v_i \neq \begin{cases} v_k; \\ \texttt{"symbol\_string"}; \end{cases}$	$6/j\vert i\vert k;$ or $6/j\vert i\,\texttt{"symbol\_string"};$
7	jump j;	$7/j;$
8	call program;	8program;
9	return;	9;
i	in_from_disk; ($v_2 \Leftarrow$ disk file v_1)	i;
o	out_to_disk; (disk file $v_1 \Leftarrow v_2$)	o;

Notice that the computer code for a step always begins with one of the digits 1 through 9, the letter i, or the letter o, the symbol indicating unambiguously the step type. If the step requires some variable to be specified, the positive integer designating the variable is always prefixed by the symbol '|', and as many '|' symbols appear for a step as there are variables to be specified. A jump step must specify the jump number j, which can be positive or negative, and here the integer prefix used is '/'. Arbitrarily long strings may represent the various integers, the end of each integer string being indicated by a nondecimal-digit symbol, such as a semicolon.

Below is a list of the 60 symbols, including a space symbol, used by the ideal computer.

```
0123456789 abcdefghijklmnopqrstuvwxyz_:;,.-+<=>|/"\(){}&^*~?
```

The only use of the backslash symbol '\' is as an "escape" symbol within a quoted string to signal that the symbol coming after the backslash should be processed without examination. This allows a quoted string to contain the quote symbol itself. For instance, `"\""` denotes '"' in quotes. The backslash itself can be specified within quotes by preceding it with another backslash. Thus `"\\"` denotes '\' in quotes.

As a simple program example, suppose we want our ideal computer to concatenate two strings and return the combined string. We then would supply as input the two strings in variables v_1 and v_2, and expect as output the concatenated string in v_1. If such a program is assigned the name example, then its representation as a string would be

```
{example>2<1;2|1|1|2;9;}
```

This example illustrates the convention that if a program takes n input strings and supplies m output strings, then its beginning code symbols are

$$\{\text{name}{>}\,n{<}\,m;\ldots$$

The symbol '>' (for input) prefixes n and the symbol '<' (for output) prefixes m. The semicolon after m is needed to separate this integer from the symbols of the program steps that follow. A terminal '}' is added after the code for the last step of the program, and the complete string, beginning and ending with a brace, is a *program segment*. If the program makes subroutine calls, then following the program segment, which we call the "main" segment, all the called program segments appear, that is, the segments corresponding to subroutines, and subroutines called by subroutines, etc. In general, a complete program with k segments has the appearance

$$\{\text{name}_1{>}\,n_1{<}\,m_1;\ldots\text{steps}\ldots\}\{\ldots\quad\ldots\}\{\text{name}_k{>}\,n_k{<}\,m_k;\ldots\text{steps}\ldots\}$$

The main segment always appears first, then all the called program segments in no particular order, except that every called program segment being used in any capacity must appear after the main segment.

For our `example` program, the step "`2|1|1|2;`" for the concatenation, and the step "`9;`" for the return, are the only steps appearing within the main segment, and there are no subroutine segments. This trivial program can be simulated on your PC, and before delving deeper into how the ideal computer does more advanced things, we explain in the next section how to use the software that comes with this book.

5.5 Viewing, Compiling, and Executing Ideal Computer Programs

The ideal computer software is designed to have no interaction with your other programs or files, to use a modest amount of your hard disk space, and to be easily removable from your PC. At some point it is possible that you may instruct the ideal computer to do some long computation and then realize that you do not have time to wait for the ideal computer's result. It may be helpful then to mention here the standard Microsoft recipe for regaining the Windows start bar: You simultaneously press the keys `Cntl`, `Alt` and `Delete`. A Windows display will appear allowing you to click on `End Task` and in this way summarily exit the ideal computer software.

We presume that you have followed the directions given in the back of this book, and now have in your Windows system the ideal computer software along with the initial ideal computer programs.

Obtain the beginning form with the title bar `Ideal Computer Disk Files` by clicking on the ideal computer name or icon in your Windows `Programs` display. This "entry" form is used for the display of program files. Click on `File` in the form's menu display to obtain a number of choices. Click on `Open` and see displayed the names of all the ideal computer programs. Making use of the scroll bar, find among these program names the `example` program mentioned in the previous section. After you click on `example`, this program's source code is displayed within the entry form.

The `example` program has its ideal computer execution code stored in the file `example`, and its source code stored in another file with the name `example.src`. The `example` execution code is brought into the entry form for inspection when you select `View Ideal Computer Code` from the `Options` menu display. The entry form again shows the source code when you select `View Source` from the `Options` menu display. Now you again see the source code that is held in the file `example.src`.

The ideal computer software allows a user to compose an ideal computer program as source code, and then convert this source code into execution code by selecting `Compile Source` from the `Options` menu display. The various ideal computer programs, whose names are displayed when you make the `File` menu selection `Open`, have been converted this way into executable programs. When the `example` source file was converted into ideal computer code, what also occurred was that an additional file with name `example.typ` was formed to record the `type` specification of the `example` source file. Every source file must specify a type with a text line that begins `type = ...`, this line appearing in the source file just before code lines. Notice that the `example` source file has the line:

```
type = string_binary_operation;
```

If we select `View Type` from the `Options` menu choices, we see displayed in the form the single line:

```
string_binary_operation
```

Thus, by choosing from the `Options` menu, in succession, `View Source`, `View Type`, and `View Code`, we see the contents of the three files that contain information about `example`.

The type files, that is, the files with suffix "`.typ`," come into play whenever you make the menu selection `Execute` or `Execute Step by Step`. We have seen that an ideal computer program receives input information in certain of its variables $v_1, \ldots, v_n$, and supplies output information in certain of its variables $v_1, \ldots, v_m$. The type specification is needed by the `Execute` software in order to prepare for the simulation by filling the program's input variables correctly, and the type is also needed after the program terminates in order to display output variables appropriately.

We are ready to test `example`. Make the `Options` menu selection `Execute`. We see that a second form appears—the "action" form—giving details about the `example` program. The action form displays the input values you specify for the ideal computer, and also displays the ideal computer's output values. Thus, for the `example` program, two input strings are requested as input, the program `example` is executed with this input, and then the output string of `example` is displayed. A stepwise execution of the "main" part of an ideal computer program is also possible, and to obtain it with `example`, make the `Options` menu selection `Execute Step by Step`. As before, two input strings must be entered, and after this is done, there is a display of the string contents of all variables used by `example`, along with the step about to be executed, the first `example` step. There is a pause now, and after you respond by hitting the `<enter>` key, the space bar, or some other key on your keyboard, the step that was displayed is performed. Afterwards there is a new display showing the next step, and again there is a pause. The two steps of `example` can be followed to program termination, whereupon, as happened before with `Execute`, the `example` program's output is displayed. If the program being simulated, unlike `example`, makes a call of a subroutine, the subroutine steps are not executed one by one but all at once, so the call subroutine step is treated as if it were a single step. You can see this, for instance, by loading in `add2natural_integers`, and executing it step by step. This program `add2natural_integers`, described in a later section, has several call subroutine steps which the `Execute Step by Step` software treats as a single step.

Before it begins the simulation of a program, the `Execute` software obtains the program's type from disk memory, and then consults the disk file `console` to find an entry for the type. The `console` type entry specifies how to prepare a program for simulation and how to handle the output of the simulated program. Load the `console` file into the entry form by making the `File` menu selection `Open` and choosing `console` from among the choices displayed. Note the line below that is at the bottom of the console file:

```
string_binary_operation;in_2strings;out_1variable;
```

A simulation of example is done by first calling the in_2strings program, then calling example, and afterwards displaying the example output by calling the out_1variable program. Thus the console type entry gives the names of other ideal computer programs that are used in the simulation.

The program example is executable but not every ideal computer program is. If a program is loaded into the entry form and the Execute menu option is disabled, this may occur either because a type for the program is missing or, if a type is present, because the type is miscellaneous or program_template, two types that are not listed in the console file.

The simulation software must assemble all the segments of any program that is be executed. You can view complete programs yourself with the help of the ideal computer program name_to_program. Choose this program in the same way as before, that is, by making the File menu selection Open and then clicking on name_to_program. When you execute this program (by the Options menu selection Execute), you are asked to supply a program name, whereupon the entire ideal computer program is displayed. For instance, type in the program name example and see its one segment program. During a simulation run, whenever a request for a program name is made, there is never any need for actually typing the name. Instead of typing, you can select the File menu option Open, and then click on any of the displayed program names. Your choice then appears as if you had typed it in. To try this now, repeat your execution of name_to_program, this time choosing example (or some other program) in the way described.

For instance, if you execute name_to_program and choose the name add2natural_integers, you see displayed eight distinct segments. The program name_to_program, like add2natural_integers, is described in more detail in a later section.

5.6 More Ideal Computer Details

The ideal computer begins operations with its various programs and subroutines stored in disk memory. Each program seqment is stored in a file having the same name as the segment. We imagine the ideal computer mechanism is supplied all the segments of some program P along with the arguments that constitute P's input, and then the mechanism begins operations. If the program P has as one of its input arguments another ideal program P_1, the program P_1 need only be identified by its name, and then

P, if it requires the segments constituting P_1, can retrieve these by calling `name_to_program`.

When the program P produces as its output another program P_2, then P need return only the name of P_2, as long as P, before terminating, stores the newly created P_2 segment in a disk file with a filename matching the P_2 name, making sure not to use a program name that is taken. That is, if a segment {`name`...} is created, then the segment must be stored in the `name` file.

We suggest a reader follow the explanations of the various programs cursorily the first time through this chapter, but actually test any program that seems interesting by executing it. Thus if the program of interest has the name `abc`, the program is chosen by making the `File` menu selection `Open`, clicking on `abc`, and then making the `Options` menu selection `Execute`. Later, to understand the program better, one can make the `Options` menu selection `Execute Step by Step`, and work through an example or two. This often makes the overall plan of the program clear. The source code that initially is displayed when a program is chosen also has comments that help explain what the program is doing.

For a Turing machine, how its starting tape is obtained is "undefined" and peripheral to the Turing machine concept. Similarly, for an ideal computer following a program P, how the input variables of P get their beginning strings is "undefined" and peripheral. We have seen that the simulation of the `example` program makes use of two other programs, `in_2strings` for loading v_1 and v_2 by means of PC console input, and `out_1variable` for the console display of v_1. Because these two programs are "peripheral," they are composed using steps of an "extended" ideal computer.

The extended ideal computer, described in greater detail in Chapter 12, has 3 additional steps for communication with a PC console, and 1 step for instantaneous cessation of computation. (These four additonal steps can be obtained by calling the four programs `in_from_console`, `out_to_console`, `cr_to_console`, and `abort`.) A program described in this chapter does not make use of these additional steps unless the program is "peripheral" to the ideal computer concept, either preparing the ideal computer for program execution, displaying the ideal computer's output, or performing some other peripheral task. There is one exception to this rule. If the input supplied to an ideal computer program is not what it is supposed to be, then what the ideal computer does next is "undefined." For instance, the program `divide2rationals` should never be given a zero divisor rational. If it were given such a divisor, it could not terminate in the usual way with a return step, as this automatically leads to some quotient string being returned to the calling program. When the input to an ideal computer program is improper, the ideal computer program is allowed to make use of the four extra steps in order to make an error report.

5.7 Programming the Ideal Computer to Add Natural Integers

Even though no ideal computer step does numeric computation of the kind our commercial computers do, nevertheless one can compose first programs for doing integer arithmetic, then programs for doing rational arithmetic, after that programs for doing interval arithmetic, and finally programs for computing with real numbers. In this section we illustrate the process by constructing a program for adding two natural integers of arbitrary length. Here we take "natural integer" to mean an unsigned positive integer or the zero integer. The program for adding two natural integers is built in stages, later programs using earlier programs through call steps.

The decimal system is used to represent our natural integers, and no prefix sign is used. We start the construction process with a simple program for incrementing by 1 a single decimal digit, the `increment_digit` program listed in Fig. 5.1. The input to this program is a single decimal digit in v_1, and the output left in v_1 is the incremented digit. Note several characteristics of the source code for this program that apply generally. A comment is always given in parentheses and is ignored when source code is converted into ideal computer execution code by the `Compile Source` software. Our source code listings always show the program name in the first line, and the second line shows the number of input arguments a program receives and the number of output arguments the program supplies on termination. The line beginning with the word "`type`" is for use by the `Execute` software, and can be ignored. Any jump step has the j jump value specified by prefixing the target step with one or more capital letters followed by a colon and then writing the same capital letters after the word `jump`. A prefix for a step may use any number of capital letters, although we usually make do with just a single capital letter. The symbol '`!=`' shown with jump steps indicates an inequality ($\neq$) comparison.

The action of this program is simple. An incoming single decimal digit is expected in variable v_1, and this variable is compared in turn with the successive strings `"0"`, `"1"`, ..., `"9"`, representing decimal digits. As soon as a match with a digit is obtained, the program output is determined and is the next higher decimal digit, except for a digit 9 match, the output in this case being 0. If the step with prefix "J:" is reached without obtaining a match, an input error is signaled and computation ceases.

A related program is the `decrement_digit` program listed in Fig. 5.2, which operates much like the previous program, except that the input decimal digit is decremented.

```
program_name = increment_digit;
number_of_arguments_in = 1; number_of_arguments_out = 1;
(increment a decimal digit (0 to 9) and return the incremented digit)
type = digit_operation;
  jump A if v1 != "0";
  v1 <- "1";
  return;
A:jump B if v1 != "1";
  v1 <- "2";
  return;
B:jump C if v1 != "2";
  v1 <- "3";
  return;
C:jump D if v1 != "3";
  v1 <- "4";
  return;
D:jump E if v1 != "4";
  v1 <- "5";
  return;
E:jump F if v1 != "5";
  v1 <- "6";
  return;
F:jump G if v1 != "6";
  v1 <- "7";
  return;
G:jump H if v1 != "7";
  v1 <- "8";
  return;
H:jump I if v1 != "8";
  v1 <- "9";
  return;
I:jump J if v1 != "9";
  v1 <- "0";
  return;
J:v1 <- "increment_digit input faulty";
  call error;
  return;
```

Figure 5.1. The program increment_digit.

```
program_name = decrement_digit;
number_of_arguments_in = 1; number_of_arguments_out = 1;
(decrement a decimal digit (0 to 9) and return the decremented digit)
type = digit_operation;
  jump A if v1 != "0";
  v1 <- "9";
  return;
A:jump B if v1 != "1";
  v1 <- "0";
  return;
B:jump C if v1 != "2";
  v1 <- "1";
  return;
C:jump D if v1 != "3";
  v1 <- "2";
  return;
D:jump E if v1 != "4";
  v1 <- "3";
  return;
E:jump F if v1 != "5";
  v1 <- "4";
  return;
F:jump G if v1 != "6";
  v1 <- "5";
  return;
G:jump H if v1 != "7";
  v1 <- "6";
  return;
H:jump I if v1 != "8";
  v1 <- "7";
  return;
I:jump J if v1 != "9";
  v1 <- "8";
  return;
J:v1 <- "decrement_digit input faulty";
  call error;
  return;
```

Figure 5.2. The program decrement_digit.

```
program_name = add2digits1carry;
number_of_arguments_in = 3; number_of_arguments_out = 2;
(add two digits and a 0-1 carry; return a sum digit and a carry)
(in_v1: digit, in_v2: digit, in_v3: carry;
out_v1: digit, out_v2: carry)
type = digit_binary_operation_with_carry;
  v4 <- "0"; (output carry stored in v4 tentatively set to 0)
  jump B if v3 = "0"; (jump if input carry is 0)
A:call increment_digit; (digit in v1 is incremented)
  jump B if v1 != "0";
  v4 <- "1"; (output carry is 1 if incremented digit is 0)
B:jump C if v2 = "0";
  (until v2 becomes 0, cyclically decrement v2 and increment v1)
  v5 <- v1; (save tentative sum digit in v5)
  v1 <- v2;
  call decrement_digit;
  v2 <- v1;
  v1 <- v5;
  jump A;
C:v2 <- v4;
  return;
```

Figure 5.3. The program add2digits1carry.

We now can use these two programs to build the program add2digits1-carry which adds two digits with a carry and supplies as output a single digit with a carry (see Fig. 5.3). Here v_1 and v_2 hold two decimal digits to be added, with v_3 holding a $0-1$ carry digit. The output the program supplies is a "sum" digit and a $0-1$ carry digit. Notice that the third step in this program calls the preceding increment_digit program, and a later step calls the decrement_digit program.

We also need the subtract2digits1carry program, which subtracts an input decimal digit and a $0-1$ input carry from another input decimal digit, with the output being a "difference" digit and a $0-1$ carry digit (see Fig. 5.4).

We now can use these last two programs to build our objective, the program add2natural_integers, which adds two natural integers in v_1 and v_2 to obtain a natural integer sum as output (see Fig. 5.5). This program works by cyclically removing terminal digits of the two input integers and adding these to form a sum digit and a carry. The sum digits are concatenated to form a nonnegative integer sum, and the process continues until all digits of the input integers have been processed.

```
program_name = subtract2digits1carry;
number_of_arguments_in = 3; number_of_arguments_out = 2;
(subtract one digit and a 0-1 carry from another digit;
return a difference digit and a carry)
(in_v1: digit, in_v2: digit subtracted, in_v3: carry subtracted;
out_v1: digit, out_v2: carry)
type = digit_binary_operation_with_carry;
  v4 <- "0"; (output carry stored in v4 tentatively set to 0)
  jump C if v3 = "0"; (jump if input carry is 0)
A:jump B if v1 != "0";
  v4 <- "1"; (output carry is 1 if digit to be decremented is 0)
B:call decrement_digit; (digit in v1 is decremented)
C:jump D if v2 = "0";
 (until v2 becomes 0, cyclically decrement v2 and decrement v1)
  v5 <- v1; (save tentative difference digit in v5)
  v1 <- v2;
  call decrement_digit;
  v2 <- v1;
  v1 <- v5;
  jump A;
D:v2 <- v4;
  return;
```

Figure 5.4. The program subtract2digits1carry.

The program add2natural_integers is the last source code program to be displayed in this book. The source code of all other programs we mention can be viewed at your computer console (see Section 5.5).

5.8 The Addition of Two General Integers

So far our ideal computer can add natural integers, but the computer also must deal with the case of general integers. Here we continue the previous notation and use positive integers without a sign prefixed, but now also allow negative integers with the minus sign prefix.

The program add2integers adds two general integers and returns the sum, another general integer. It makes use of the program add2natural_integers and another program subtract2natural_integers, which we describe first.

The program subtract2natural_integers is somewhat more complicated than add2natural_integers because, unlike that program, subtract2-

```
program_name = add2natural_integers;
number_of_arguments_in = 2; number_of_arguments_out = 1;
(add two natural integers and return the natural integer result)
type = natural_integer_binary_operation;
  v5 <- v1; (save first input integer in v5)
  v6 <- v2; (save second input integer in v6)
  v3 <- "0"; (initial carry is 0)
A:v1 <- v5]; (remove last digit of first input)
  v2 <- v6]; (remove last digit of second input)
  call add2digits1carry;
  v4 <- v1 + v4; (prefix sum digit to digits already formed)
  v3 <- v2;   (output carry becomes input carry)
  jump B if v5 != "";
  v5 <- "0";  (v5 set to 0 in case v6 still has digits)
  jump A if v6 != "";
  v6 <- "0";  (v6 set to 0 in case carry = 1)
  jump A if v3 = "1"; (jump if last carry is 1)
  v1 <- v4;   (all digits processed when this step is reached)
  return;
B:jump A if v6 != "";
  v6 <- "0";
  jump A;
```

Figure 5.5. The program add2natural_integers.

natural_integers sometimes must return a negative integer. It is convenient to first construct the program subtract2natural_integers_signal_minus, which is similar in form to add2natural_integers.

The preparatory program subtract2natural_integers_signal_minus returns a natural integer if the difference of its two input natural integers is nonnegative, and should this not be the case, the program returns the minus sign symbol alone to signal this. The program subtract2natural_integers calls the preparatory program, supplying this program with its own input, and if the minus sign signal is returned, subtract2natural_integers obtains the correct result by calling the preparatory program a second time but with the input integers reversed, and then prefixing a minus sign to the natural integer returned.

With both programs add2natural_integers and subtract2natural_integers available, it is now easy to construct the program add2integers that adds two general integers. This program examines each input integer to determine whether it has a minus sign prefix, deleting this sign if it is present, and then according to the various cases that arise, calls either add2natural_integers or subtract2natural_integers, sometimes prefixing a minus sign to the obtained result when this is required.

5.9 The Subtraction, Multiplication, and Division of Integers

The program for doing general integer subtraction, subtract2integers, is similar to the program for doing general integer addition. Like add2integers, the program subtract2integers first examines its two input integers to determine whether they are negative, deleting any minus sign prefixes in the process, and then depending on whether no minus sign, one minus sign, or two minus signs are encountered, calls either add2natural_integers or subtract2natural_integers, with the sign-deleted integers as their input, and then gives the subroutine's output as its own output, sometimes first prefixing a minus sign.

With multiplication, it is convenient to first construct the program multiply2natural_integers, which obtains the product of two natural integers. This program uses the standard algorithm, taught in elementary school, to obtain the product. For instance, if 234 is the multiplier, and 5555 is the multiplicand, we need the products of 5555 by the digits 4, 3, and 2. These three single-digit products can be obtained by repeated additions of 5555. We add each single-digit product to a progressive sum, which initially is zero, and then remove the last digit of the sum and prefix it to a progressively growing string of terminal product digits. This cycle is repeated until all multiplier digits have been processed. The multiplication result is obtained by prefixing the final sum to the final string of terminal digits.

The program multiply2integers multiplies two general integers by first deleting any prefix minus signs of its operands and then calling multiply2-natural_integers as a subroutine, supplying it with the obtained natural integer operands. In certain cases a minus sign is prefixed to the subroutine's result. If the two input integers have just one minus sign prefix between them, the subroutine result has a minus sign prefixed unless the result is zero, while if the two input integers both have minus sign prefixes, the subroutine result is unchanged.

The program divide_natural_integer_by_positive_integer divides a positive divisor integer into a natural dividend integer, and obtains a natural integer quotient along with a natural integer remainder. The standard division algorithm is used here. First the digits appearing in the divisor are counted, and likewise the digits appearing in the dividend. If there are fewer dividend digits than divisor digits, the quotient is zero and the remainder is the dividend.

If the number of dividend digits is at least as large as the number of divisor digits, then the procedure is much like doing a multiplication in reverse. The dividend is split into two parts, a leading part with the same number of digits as the divisor, and a trailing part containing the rest of the dividend digits. The divisor is repeatedly subtracted from the leading part until the result becomes negative, and by counting the number of subtractions up to but not including the one causing a negative result, one digit of the quotient is obtained. The leading part now is restored to the value it had prior to the subtraction causing the negative result. Then the first digit of the trailing part is appended to the leading part, and the whole process is repeated to obtain another quotient digit. This continues until the trailing part, being empty, can no longer contribute a digit to the leading part. This indicates that all quotient digits have been obtained, and the quotient is formed as the string of all the individual quotient digits. The remainder is the final restored value of the leading part. This process sometimes leads to the quotient having an unneeded initial zero digit, which must be removed when present.

5.10 Rational Number Arithmetic

Now that we have a battery of programs for doing integer arithmetic, we can consider how to compute with rational numbers. Our first step is to obtain for each rational number a unique symbol representation. Suppose the two integers p and q define the rational number p/q. The integers p and q must have any common positive divisor greater than 1 divided out, and q must be positive. After p and q are brought to this required form, the rational is represented by a string that has first the symbols for p, then the separating symbol '/', then the symbols for q. This string gives the *standard rational string* for any rational number. A negative rational number can be detected by examining just the first symbol of the string, which is a minus sign whenever p is negative. In this system, the rational number zero always appears as the symbol string "`0/1`." In addition, it is easy to determine whether two rationals are equal, this occurring if and only if the two strings representing the rationals are identical.

The rational operations $+$, $-$, $\times$, and $\div$ are listed as equations at the beginning of Section 3.3. To execute one of the four operations on two input rational strings, it is only necessary to split the symbol string for each operand r into two strings representing the integers p and q used by r, then combine the resultant four integers according to the rules listed in Section 3.3, and in this way obtain two integers p' and q' defining the rational number result. From these two integers the result's standard rational string is obtained.

The four programs `add2rationals`, `subtract2rationals`, `multiply2-rationals`, and `divide2rationals` do the four rational operations, and all these programs are executable. When entering a rational number at the computer console as an input for these programs, specify the rational number in the form p/q using any integer for p and any nonzero integer for q, specify the rational as an integer i, or specify it in the decimal form $d_0.d_1d_2 \ldots d_k$. Your console entries are converted into standard rational strings and supplied as variables v_1 and v_2 to these programs.

These four rational operation programs, besides calling the various programs already described for doing integer operations, call two other programs, `rational_to_integer_and_positive_integer` and `integer_and_non-zero_integer_to_rational`. The first-named program converts a standard rational string into the numerator p string and the denominator q string by locating the separating symbol '/'.

The second program is more complicated, and converts its two input integers p, q, with $q \neq 0$, into a standard rational string. If q is a negative integer, the program changes it to a positive integer by removing its prefix sign and changing p appropriately, and after this examination of p and q, the program calls on still another program, `integer_and_positive_integer_to_rational`, to complete the conversion of the p, q pair into a standard rational string. We describe this third program next.

The integers p and q may have a common divisor that needs to be removed. The greatest common divisor of $|p|$ and q can be obtained by the process known as Euclid's method or Euclid's algorithm. The program `greatest_common_divisor_of_2positive_integers` uses this process on any two positive integers supplied to it in order to return the greatest common divisor. The program `integer_and_positive_integer_to_rational` obtains the greatest common divisor of p and q by calling the program just described, divides out the common divisor if it is >1, and then is able to construct the standard rational string.

Euclid's method, applied to any two positive integers p_1 and p_2, starts by dividing p_2 into p_1, obtaining the quotient integer q_1 and remainder integer p_3:

$$p_1 = q_1 p_2 + p_3$$

Because $p_3 = p_1 - q_1 p_2$, any common divisor of p_1 and p_2 is also a divisor of p_3, and hence a common divisor of p_2 and p_3. Moreover, the preceding displayed equation shows that any common divisor of p_2 and p_3 is also a divisor of p_1, and hence a common divisor of p_1 and p_2. Therefore, we can replace the problem of finding the greatest common divisor of p_1 and p_2 by the problem of finding the greatest common divisor of p_2 and p_3. Euclid's method consists of repeatedly replacing one pair of integers by another pair of integers, all integer pairs having the same greatest common divisor, until finally the greatest common divisor becomes obvious. In general, a succession of pairs $(p_1, p_2), (p_2, p_3), (p_3, p_4), \ldots,$ is obtained, with the numbers $p_3, p_4, \ldots,$ being the successive division remainders.

If p_1 is smaller than p_2, then $q_1 = 0$ and $p_3 = p_1$, and the replacement process simply reverses the order of the two integers. Even in this case, however, notice that the remainder integer is less than the divisor integer it replaces. In the next cycle, the remainder is smaller still, and eventually, we must obtain a remainder of zero. When this happens, the greatest common divisor is revealed as equal to the integer paired with the zero.

You can follow Euclid's method with various examples of your choice by stepwise execution of `greatest_common_divisor_of_2positive_integers`. This program does the divisions required by Euclid's method by calling `divide_natural_integer_by_positive_integer`, a program described in the preceding section.

There are three other programs that do rational number arithmetic. These are the programs `maximum_of_2rationals`, `minimum_of_2rationals`, and `absolute_value_of_rational`.

5.11 Interval Arithmetic

Now that we have programs for executing the four operations $+$, $-$, $\times$, and $\div$ on rational numbers, we can consider doing interval arithmetic. Here we have rationals for the approximant m and error bound e of an interval $m \pm e$. We adopt the convention that an interval $m \pm e$ is to be represented by a symbol string consisting of a standard rational string for m followed by the separation symbols `+/-` and then the standard rational string for e, the complete string being called the *standard interval string*. The minus sign of the separator `+/-` is always followed by a decimal digit because the error bound e is always positive or zero.

The interval arithmetic equations (2.2)–(2.5) listed in Section 2.3 are easily implemented, but recall that the division operation (2.5) was revised in Section 2.4 to yield the interval $0 \pm \infty$ for the divide error case. We need a representation for $0 \pm \infty$, and now take this to be the string `0/1+/-1/0`. Here we note that the error bound `1/0` is not a rational number because the quotient integer is zero. The four programs `add2intervals`, `subtract2intervals`, multiply2intervals, and divide2intervals perform the four interval arithmetic operations. Each of these four programs, by calling the program interval_to_rational_and_nonnegative_rational, decomposes both input intervals into a rational approximant and a rational error bound, then with the obtained four standard rational strings, constructs the two rational parts of the result interval, whereupon the program rational_and_nonnegative_rational_to_interval is called to combine these parts into a standard interval string. Two of the four programs, multiply2intervals and divide2intervals, need to form absolute values of rational numbers, and for this they call on the program absolute_value_of_rational, which is a simple program that deletes the minus sign prefix of a rational number if it is present. In Section 2.4 all four interval arithmetic operations were revised to yield $0 \pm \infty$ if either operand is $0 \pm \infty$, and these four programs adhere to this revision.

Besides these four basic interval arithmetic programs, there are three others: maximum_of_2intervals, minimum_of_2intervals, and absolute_value_of_interval. These programs are similar to the ones already described and make use of rational operation programs mentioned in the previous section.

5.12 The Retrieval of a Program from Disk Memory

Programs for integers, rationals, and intervals are simpler than programs for real numbers, because an integer, rational, or interval can be defined by a string of symbols, whereas a real number is defined by a program, specifically, an approximation algorithm. Thus, when we add or multiply real numbers, our operands are programs, and our operation result is another program. Our ideal computer is postulated with an ideal disk memory to make dealing with real numbers more convenient. We can identify a real number by its program name, and if the actual program is needed, it can be retrieved from disk memory.

Our postulated ideal computer begins operations by receiving, in some unspecified way, all the segments of some program, and being supplied, again in some unspecified way, with the contents of all input variables used by the program. The ideal computer, now having available all program steps and the input information, can proceed to perform these steps. If eventually the ideal computer encounters a return step in the program's main segment, then the contents of all variables with output values remain behind, and the ideal computer ceases computation. In Section 5.22 we describe a way of doing all the things required of our ideal computer, and the description serves to show that our ideal computer is a Turing machine.

The programs examined so far have been converted to ideal computer code by the supplied `Compile Source` software routine, which itself is not an ideal computer program, but may be considered an extraneous device for obtaining ideal computer code and storing it in the ideal computer's disk memory. The system of storing the main segment execution code of a program with name `abc` in the ideal computer's disk memory under the file name `abc` is merely one possible way of making use of the ideal computer's disk memory. This system has the advantage of avoiding repetitions of ideal computer code in disk memory. If complete ideal computer programs were stored instead, frequently used subroutines would be recorded in many places.

Because sometimes there is a need to assemble a complete program from merely a program name, the system of storing individual segments requires an ideal computer program that collects the separate parts of a given program. The program that does this, encountered earlier, is `name_to_program`. This program retrieves complete programs by using two work lists containing program names, L(found) and L(get), and a list of segments L_P. Initially, all three lists are empty; when `name_to_program` terminates, the list L_P holds all the segments of the named program.

The program `name_to_program` takes the input program name, obtains from disk memory the corresponding main segment, loads L_P with it, and adds the input program name to L(found). Then it scans the obtained main segment to obtain a list of called subroutine names (by calling `extract_names_from_segment`), and adds these to the list L(get). The list L(get) is a list of program names whose segments are to be retrieved from disk storage, and the list L(found) is a list of program names whose segments have already been retrieved.

Cyclically, `name_to_program` removes the first name from L(get) and checks L(found) to determine if the removed name is present (by calling `check_list`). If the name is present, nothing further is done this cycle; if the name is not present, the name is added to L(found), the program's segment is retrieved and appended to L_P, and then the segment is scanned for program names appearing in call steps, with the list of names obtained appended to the list L(get). These cycles continue until L(get) is empty, whereupon the list L_P can be returned as the desired complete program.

The program `name_to_subroutine_list` behaves just like `name_to_program` except that it returns the list L(found) instead of L_P. When we execute `name_to_subroutine_list` and supply some program name, say `abc`, we obtain the list of all subroutines used by `abc`. The list begins with `abc`, and the subroutine names following are separated from each other by a semicolon.

5.13 The Real Number π

In Section 2.1 we displayed the various values returned by an approximation algorithm of type E for the real number π, and in this section we describe the ideal computer program returning these values. There are many ways to compute approximations to π, and we make use of the following equation for $\pi/4$, Eq. (351) in Jolley's book listing infinite series [18].

$$\frac{\pi}{4} = \tan^{-1}\frac{1}{2} + \tan^{-1}\frac{1}{3}$$

$$\frac{\pi}{4} = \left(\frac{1}{2} - \frac{1}{3 \cdot 2^3} + \frac{1}{5 \cdot 2^5} - \dots\right) + \left(\frac{1}{3} - \frac{1}{3 \cdot 3^3} + \frac{1}{5 \cdot 3^5} - \dots\right) \tag{5.1}$$

This gives $\pi/4$ as a sum of two alternating infinite series, each of the series obtained from the Maclaurin series for $\tan^{-1} x$:

$$\tan^{-1} x = x - \frac{x^3}{3} + \frac{x^5}{5} - \dots$$

We can verify that the two angles $\tan^{-1}\frac{1}{2}$ and $\tan^{-1}\frac{1}{3}$ sum to $\pi/4$ by using the formula for the tangent of the sum of two angles:

$$\tan(\alpha + \beta) = \frac{\tan\alpha + \tan\beta}{1 - \tan\alpha\tan\beta}$$

$$\tan\left(\tan^{-1}\frac{1}{2} + \tan^{-1}\frac{1}{3}\right) = \frac{\frac{1}{2} + \frac{1}{3}}{1 - \frac{1}{2}\cdot\frac{1}{3}} = \frac{\frac{5}{6}}{\frac{5}{6}} = 1 = \tan\frac{\pi}{4}$$

Our ideal computer program `pi_e` for computing $\pi(E)$ obtains approximations to π by using the alternating infinite series shown next, a rearrangement of Eq. (5.1), the series terms decreasing in magnitude and tending toward zero:

$$\pi = 4\left[\frac{1}{1}\left(\frac{1}{2} + \frac{1}{3}\right) - \frac{1}{3}\left(\frac{1}{2^3} + \frac{1}{3^3}\right) + \frac{1}{5}\left(\frac{1}{2^5} + \frac{1}{3^5}\right) - \dots\right]$$

The sum of such an alternating series to a particular number of terms differs from the series limit by no more than the magnitude of the first term not summed. The $\pi(E)$ program sums these terms until the magnitude of the term next to be summed is within the E bound, at which point an approximation interval can be returned. The interval approximant is the series sum so far accumulated, and the error bound is the magnitude of the final term that was computed but not summed.

If you load the `pi_e` program and execute it, you can verify that its output matches that shown for $\pi(E)$ in Section 2.1.

Because we use three different approximation algorithms for real numbers, of types E, D, and the main type K, we differentiate real number program names by the two-symbol endings _e, _d, or _k. This is why the $\pi(E)$ program has the name pi_e.

The source code for the pi_e program has the line "type = real_e;" just before the code lines. The types of approximation algorithm programs are real_e, real_d, or real_k, respectively, according to whether the algorithms are of type E, D, or K.

5.14 Changing an Approximation Algorithm

The E approximation algorithm for π can be changed into a D approximation algorithm by the program real_type_e_to_real_type_d. This program takes the name of an E algorithm program, generates the corresponding D algorithm program, stores the new program segment in disk memory, and then returns the name of the newly created program.

If the ideal computer program abc creates another program Q, it can do this by assembling in its variables code for Q's steps. However, if the Q program has more than one or two steps, then reading the abc source code often does not clarify what Q will do. We can make the Q code assembled by abc more understandable by describing it in general terms in the file abc_, with the file abc_ capable of being compiled into ideal computer code. This makes the abc program simpler, for now abc just retrieves the abc_ code from disk memory and makes simple substitutions in it to obtain the desired target program Q. The abc source file explains what abc does, and the abc_ source file explains what Q does.

This comment applies to the program real_type_e_to_real_type_d. The Q program that is created here has its general form shown as the program real_type_e_to_real_type_d_. If you load this second program, you can read the source code and also examine the ideal computer code. Here note that name1 serves as a placeholder for the actual name of the E algorithm supplied as input. The type of any program whose name ends in the underline symbol '_' is always program_template. Such programs are not executable, that is, when this program type is loaded, the Options menu selection Execute is disabled. The program real_type_e_to_real_type_d replaces name1 in the template program with the E algorithm input name, and in this way creates a program that calls the E algorithm and changes the obtained rational interval into a decimal interval.

An ideal computer program P which creates another ideal computer program Q returns only the Q program name, after first depositing in disk memory the new code segment created. The program P must avoid using a program name that already is taken. The task of choosing program names is handled by the subroutine `record`, which takes a supplied code segment, gives the code segment an arbitrary unused program name, deposits the now renamed code segment in the disk file with the new name, and then returns the new program name. The program P calls `record` and returns whatever name `record` supplies.

Program names must begin with a letter, and each of the other symbols of the name can be a letter, a decimal digit, or the symbol '_'. The method that the program `record` uses to obtain distinctive program names is to append a serial number to the one letter prefix `m`, adding 1 to the previous serial number each time it is called to assign a name. Thus the series of names `record` generates is `m1`, `m2`, `m3`, ...

The $\pi(E)$ algorithm may be converted into a D algorithm by executing the program `real_type_e_to_real_type_d`. In response to the request for a `real_e` number, either type the name `pi_e` or choose the `File` menu option `Open` and click on `pi_e`. The name of the newly constructed D algorithm is returned. Whatever name is returned will now be found among the list of program names shown when you make the `File` menu choice `Open`. If, say, the displayed new name is `m34`, you now can execute `m34` and obtain output for various D values you supply.

The conversion of a D algorithm into a K algorithm is done by the program `real_type_d_to_real_type_k`, and the conversion of a K algorithm into an E algorithm is done by the program `real_type_k_to_real_type_e`. These three conversion programs are then the means for converting one type of approximation algorithm into another type. Thus one can convert the $\pi(D)$ algorithm into a $\pi(K)$ algorithm, then convert the $\pi(K)$ algorithm back into a $\pi(E)$ algorithm that will differ somewhat from the starting $\pi(E)$ algorithm. The conversion of a K algorithm to a D algorithm is a frequent need, so the program `real_type_k_to_real_type_d` is also provided; this program first calls `real_type_k_to_real_type_e` and then calls `real_type_e_to_real_type_d`.

You can change an `m`-number name to a more mnemonic name by executing `rename_program`. This program takes the program name you enter, obtains the program segment, then takes your entered new name and first checks that the name is not already in use. If the name is free, `rename_program` replaces the name that immediately follows the segment's initial brace with the new name, and stores the changed segment in the disk file with the new name. The `rename_program` also copies the old-name type file into the new-name type file, because this information is needed when you execute the new name.

The two counterparts to the `pi_e` program, that is, the D- and K-type π approximation algorithms, are among the programs supplied with the CD. These programs have the names `pi_d` and `pi_k`, and do not have corresponding source code files because the programs were created by using the conversion programs and `rename_program` in the way just described.

5.15 Rational Numbers Converted to Real Numbers

A real number is defined by a K approximation algorithm, so a rational number, when viewed as a real number, must have a corresponding approximation algorithm. The program `rational_to_real_type_k` takes an input rational number, constructs and stores the corresponding simple K approximation algorithm, and returns the `m`-number name assigned. Such rational K approximation algorithms were described in Section 2.2.

It makes an interesting exercise to convert some rational number into a K approximation algorithm, then convert the K algorithm into the D form by executing `real_type_k_to_real_type_d`. With the D algorithm, one can obtain decimal approximations to the rational to an arbitrary number of decimal places.

The four programs `zero_k`, `one_k`, `minus_one_k`, and `two_k` provide K algorithms for the four rational numbers 0, 1, -1, and 2. These approximation algorithms do not have source code files because each was generated by executing the `rational_to_real_type_k` program and then executing the `rename_program`. (If some other rational number, say 10, is needed, a program `ten_k` can be generated the same way.) There are also the four programs `zero_e`, `one_e`, `minus_one_e`, and `two_e`, giving E algorithms for these four key rationals, and the four programs `zero_d`, `one_d`, `minus_one_d`, and `two_d`, giving D algorithms. All these programs have no source code files because they were generated by executing the various conversion programs and the `rename_program`.

5.16 *n*th Roots of Real Numbers

The procedure described in Section 4.4, after being made more efficient, is used by the program `nth_root_of_real_type_d` to generate a D algorithm for $\sqrt[n]{a}$ from the D algorithm for a. For instance, to generate a $\sqrt{2}(D)$ algorithm, one executes `nth_root_of_real_type_d` and enters `two_d` in response to the request for a `real_d` program name. One then obtains an `m`-number name for the generated D algorithm. Because $\sqrt{2}$ is often needed for real number computations, a program to generate it, `sqrt2_d`, is included among the CD programs. It was formed by using the nth root program and then renaming the resultant `m`-number program. There is also a `sqrt2_k` and a `sqrt2_e` program, created by using the various conversion programs and the renaming program in the manner previously described.

Recall that the procedure described in Section 4.4 for finding $\sqrt[n]{a}(D)$ begins by obtaining $a(nD)$. If n is odd, any value for $a(nD)$ is acceptable, but if n is even and $a(nD)$ is a negative interval, this indicates an error in the choice of the real number a. Because of this difference between the even-n and odd-n cases, the program `nth_root_of_real_type_d` first determines the parity of n, and then calls either an even-n root program or an odd-n root program. The even-n root program, after checking that the $a(nD)$ interval is not negative, converts $a(nD)$ to a natural integer m by deleting the decimal point and the terminal symbols `+/-1`, and afterwards stripping away possible unneeded initial zero digits. (In this conversion of the interval $a(nD) = d_0.d_1 \ldots d_{nD} \pm 1$ to an integer m, there is at least one beginning 0 digit discarded whenever d_0 is 0.) Then digit by digit, a natural integer q is constructed, whose nth power is $\leq m$, while the nth power of $q + 1$ is $> m$. The integer q then has a decimal point inserted D places back from its last digit, the `+/-1` string is reattached, and this interval is returned. The odd n root program is similar, except that if $a(nD)$ has a minus sign, the sign is deleted, the process just described is carried through, and then the sign is reattached to the result.

5.17 Arithmetic Operations on Real Numbers

The four programs `add2reals_type_k`, `subtract2reals_type_k`, `multiply2-reals_type_k`, and `divide2reals_type_k` perform the four operations $+$, $-$, $\times$, and $\div$ on real numbers. When you execute one of these programs, you are asked to specify by name two K approximation algorithms (type `real_k`), and the name of the newly created K algorithm is returned. The `real_k` names that can be entered are those of the CD-supplied `real_k` programs, which are `pi_k`, `sqrt2_k`, `one_k`, `minus_one_k`, `zero_k`, and `two_k`; also possible is any `m`-number name of a created `real_k` number. The returned program can then be executed to obtain a series of rational interval approximations. The K algorithm also can be converted to the more convenient D algorithm giving decimal intervals, by executing the appropriate conversion program.

Each of the four programs for addition, subtraction, multiplication, and division of real numbers first attempts to determine whether the two input operands are rational. There is no general method of doing this, but it is possible to determine whether an operand program has the same steps as do the programs generated by `rational_to_real_type_k`. The test for operand rationality thus is a check for one specific program type. If both operands are rationals, then the two rationals are combined using rational arithmetic, and the rational result is converted into a `real_k` program by calling `rational_to_real_type_k`. The obtained `real_k` program then is a simple one in this case. If either operand program fails the rationality test that is made, then the method described in Section 3.4 is used to generate a real_k program. This general method of real number computation uses the two operand programs as subroutines.

There are three other programs for doing real number arithmetic: maximum_of_2reals_type_k, minimum_of_2reals_type_k, and absolute_value_of_real_type_k. These programs are similar in structure to the programs just described.

5.18 Rational Sequence Arithmetic

In general, when two real numbers are combined with a rational operation, an approximation algorithm is created that calls the two operand algorithms as subroutines. If we discard the operand algorithms, then we also lose the result algorithm. Computation with rationals does not entail such interconnectivity, and it is easier to demonstrate sequence arithmetic if we restrict ourselves to *rational sequences*, that is, sequences taking rational values only.

The identity_rational_sequence program behaves like a rational sequence a_n whose nth term is n. For any input positive integer n, this program simply converts the integer n to the rational $n/1$.

A constant rational sequence a_n is a sequence whose nth term equals an unchanging rational number r. The program create_constant_rational_sequence takes an input rational number and returns the corresponding constant rational sequence program.

A number of programs combine rational sequences in various ways to form new sequences. Rational sequences can be combined rationally by means of the four programs add2rational_sequences, subtract2rational_sequences, multiply2rational_sequences, and divide2rational_sequences. Each of these programs takes two sequences a_n and b_n and constructs the sequence c_n equal to $a_n \circ b_n$, where $\circ$ is one of the four rational operations. By using all six of the rational sequence programs so far described, one can construct any given rational function of n.

The program rational_sequence_sum constructs from an input sequence a_n the sequence $s_n = \sum_{i=1}^{n} a_i$, and the program rational_sequence_product constructs from a_n the sequence $p_n = \prod_{i=1}^{n} a_i$. Using these two programs, if we specify a_n to be the identity rational sequence, we obtain the sequences $s_n = n(n+1)/2$ and $p_n = n!$. The sequence of Section 4.1, having every rational as a term, is generated by the program rational_sequence_with_every_rational_a_sequence_term.

Each of these rational sequence programs can be converted to general real-valued sequence form by making small changes in the source code.

For instance, the `identity_rational_sequence` program would need before its return step the step `call rational_to_real_type_k`. The template program `add2rational_sequences_` would need the step `call add2rationals` changed to `call add2reals_type_k`.

5.19 Function Programs

The program `identity_function` models the function $f(x) = x$, and for any input x returns x as output. This simple program has just one step, a return. The program `create_constant_function` takes as input a `real_k` number c and returns the corresponding constant function $f(x) = c$.

Functions can be rationally combined by means of the programs `add2-functions`, `subtract2functions`, `multiply2functions`, and `divide2-functions`. These four programs take two input functions $f(x)$ and $g(x)$ and return the function $f(x) \circ g(x)$, where $\circ$ is one of the four rational operations $+$, $-$, $\times$, $\div$. The programs `maximum_of_2functions` and `minimum_of_2functions` take two input functions $f(x)$ and $g(x)$ and return the functions $\max(f(x), g(x))$ and $\min(f(x), g(x))$, respectively. The program `absolute_value_of_function` takes just one input function $f(x)$ and returns the function $|f(x)|$.

The program `nth_root_of_function` takes an input function $f(x)$ and an input positive integer n and returns the function $\sqrt[n]{f(x)}$. If the input function is the identity function, and the integer n chosen is 2, then the function created is $\sqrt{x}$. With a created $\sqrt{x}$ function it is possible to explore the phenomenon described in Section 4.7. For instance, if the rational number -0.00000001 is converted into a real number by executing the program `rational_to_real_type_k`, then when this `real_k` number is supplied to $\sqrt{x}$, the error is not detected and $\sqrt{x}$ returns a function value, that is, a supposed K algorithm. This faulty K algorithm returns an interval for $K = 1, 2, 3, 4$, and only when K is 5 or more, is the error in the x argument detected.

A polynomial function can be created by generating the proper constant functions and combining these appropriately with the identity function. For instance, if we want the polynomial $P(x) = x^2 + 2x + 3$, we generate with `create_constant_function` the functions $f_1(x) = 2$ and $f_2(x) = 3$, and then can obtain $P(x)$ as $x \cdot (x + f_1(x)) + f_2(x)$. A more convenient way of building a polynomial function is provided by the two programs `create_finite_sequence` and `create_polynomial_function`. The first program creates a finite sequence a_n that is defined for $n = 0, 1, \ldots,$ up to a specified highest index N. The second program takes as input any such finite sequence a_n and creates the polynomial $\sum_{n=0}^{N} a_n x^n$.

5.20 The Exponential Function e^x

When we supply a polynomial function an argument $x(K)$, the function value approximation algorithm $y(K)$ generates the polynomial from the polynomial coefficents and $x(K)$, using appropriate multiplication, addition and subtraction operations. As K increases, the error bounds of the initial operands decrease in the way required, so the error bound of $y(K)$ decreases similarly.

Consider a more complicated case, the function e^x, which can be generated as an infinite series that converges for all x:

$$e^x = 1 + x + \frac{x^2}{2!} + \frac{x^3}{3!} + \ldots$$

Here the function value $y(K)$ can be computed as a partial sum of the series, that is, a Taylor polynomial, if we arrange to increase the degree of the Taylor polynomial in some fashion as K increases, and also arrange to account for the series truncation error. The program `exponential_function` takes $y(K)$ to be $\sum_{n=0}^{N} x(K)^n/n!$, computed by interval arithmetic, with N being chosen large enough so that the interval $x(K)/N$ is inside the interval $[-\frac{1}{2}, \frac{1}{2}]$. Then the truncation error is bounded by the magnitude of the last term of the partial sum, because we have

$$\begin{aligned}
\text{truncation error} &= \left| \frac{x^{N+1}}{(N+1)!} + \frac{x^{N+2}}{(N+2)!} + \frac{x^{N+3}}{(N+3)!} + \ldots \right| \\
&\leq \frac{|x|^{N+1}}{(N+1)!} + \frac{|x|^{N+2}}{(N+2)!} + \frac{|x|^{N+3}}{(N+3)!} + \ldots \\
&= \frac{|x|^N}{N!} \left(\frac{|x|}{N+1} + \frac{|x|^2}{(N+1)(N+2)} \right. \\
&\quad \left. + \frac{|x|^3}{(N+1)(N+2)(N+3)} + \ldots \right) \\
&< \frac{|x|^N}{N!} \left(\frac{|x|}{N} + \frac{|x|^2}{N^2} + \frac{|x|^3}{N^3} + \ldots \right) \\
&\leq \frac{|x|^N}{N!} \left(\frac{1}{2} + \frac{1}{2^2} + \frac{1}{2^3} + \ldots \right) \\
&= \frac{|x|^N}{N!} \left(\frac{1/2}{1 - 1/2} \right) \\
&= \frac{|x|^N}{N!}
\end{aligned}$$

The program `exponential_function` adds a truncation error bound to the error bound obtained in summing the series with interval arithmetic, to obtain an approximation interval for the exponential function. If we execute this program using for x the argument one_k, and convert the obtained real_k function value to real_d form by executing real_type_k_to_real_type_d, from it we can obtain as many decimal places for Euler's constant e as we are willing to wait for.

5.21 Two Semifunction Examples

Suppose we want a function $f(x)$ that for any x returns a value equal to the greatest integer $\leq x$. In computable calculus such a function is not possible because of the many argument points where the function is discontinuous. The program semifunction1 attempts to realize this greatest integer function in a straightforward way, and returns an integer for every input approximation algorithm. Because the greatest integer function has only integer values, semifunction1 returns an integer instead of an approximation algorithm, the usual output of a function. (If an approximation algorithm is required, perhaps to serve as an input to another function, the returned integer string can be converted to a standard rational string by appending the symbols "/1," then converting to a real_k by calling rational_to_real_type_k.) The input to semifunction1 is an E approximation algorithm instead of the usual K algorithm, because the computation method requires this algorithm type.

The program semifunction1 obtains an interval $x(E)$ with E taken as $1/100,000$, and then calls interval_to_integer and returns whatever integer is supplied. Here the called subroutine ignores the interval error bound, and calculates the greatest integer less than or equal to the interval approximant. Thus, semifunction1 returns the correct value whenever the $x(1/100,000)$ interval is a point, which occurs if $x(E)$ is the E form of a K algorithm created by rational_to_real_type_k. This occurs, for instance, if $x(E)$ is generated by one_e, two_e, zero_e, or minus_one_e. However, rational number approximation algorithms sometimes do not return point intervals, and for such algorithms semifunction1 could be in error, implying that the program really is a semifunction and not a function.

For instance, suppose the rational number 1 is calculated by using the geometric series

$$1 = \frac{1}{2} + \frac{1}{4} + \frac{1}{8} + \ldots + \frac{1}{2^n} + \ldots$$

The program one_by_geometric_series_e makes such a series computation to obtain E approximations to 1. If this E algorithm is supplied to semifunction1, the program returns 0, the wrong integer value. We can improve semifunction1 by decreasing the rational E value used to obtain the approximation interval, and perhaps by trying to make an internal examination of the E algorithm supplied. Regardless of how we improve semifunction1, the next chapter will show that our program can never realize its target function.

When semifunction1 is executed, this is done with two separate simulation runs. The semifunction1 program creates another program to do the actual computation, and this second program is executed after semifunction1 terminates. When a program P is supplied the name of another program as an input, it cannot call this second program as a subroutine. In order to do this, P would have to insert a call step with the new name into one of its own segments. We have postulated our ideal computer as a mechanism that follows a program supplied to it as a string of segments, and this plan is in fact carried out by the Execute software. A program can change the various segments of its own program stored in disk memory, but this would not affect the program string being followed by the postulated mechanism (or by Execute). Nevertheless, a program can obtain needed output values of a second program designated by name by calling an appropriate subroutine. In the next section we describe the simulator program execute_program that simulates the steps of whatever program is specified to it by name. The needed value $x(1/100,000)$ could be obtained by calling execute_program, and the program semifunction1a makes the same calculation that semifunction1 does in just this way. The program semifunction1a is somewhat simpler than semifunction1, but it is also much slower, because when execute_program simulates an ideal computer step, it needs several hundreds of its own steps to do it. If you execute semifunction1a, you will be able to get an output without too much wait if your input is one_e or two_e. Be prepared for a long wait if you choose pi_e or sqrt2_e as your input. The program semifunction1 can handle all these inputs reasonably quickly. The two run execution of semifunction1 is a concession to practicality. If we were able to simulate the ideal computer about a thousand times faster, there would be no need of employing this two run system.

A second semifunction is obtained when we attempt to approximate the discontinuous function $\operatorname{sgn} x$. The program semifunction2, given an input $x(E)$ algorithm, obtains $x(1/100,000)$, and according to whether the rational interval approximant is positive, negative, or zero, returns 1, -1, or 0. The program semifunction2 is executed in two separate simulation runs, just as was described for semifunction1. (The alternate version provided, semifunction2a, corresponds to the program semifunction1a and is executed in one run.) Again we have a semifunction and not a function, because the returned integers are not consistent. The zero_by_geometric_series_e program is an E algorithm for zero, calculated by using the series

$$0 = 1 - \frac{1}{2} - \frac{1}{4} - \frac{1}{8} - \dots - \frac{1}{2^n} - \dots$$

When this E algorithm is supplied to `semifunction2`, we obtain the integer 1 instead of 0. As with `semifunction1`, there are many ways to improve our program, but no matter what we do, we can never succeed in making it a function.

5.22 The Beginning of the Nonsolvability Proof for Problem 3.1

The rest of this chapter is devoted to filling in the details of the nonsolvability proof of Section 3.4, which shows that there can never be a program for determining whether any real number is zero. The previous sections of this chapter have shown how to convert various numerical procedures into ideal computer programs, and it should be clear now that there is no essential difficulty in converting any concrete procedure into a corresponding ideal computer program.

The key to the sketched proof of Section 3.4 is the subroutine $P_S(P_1, P_2, n)$, which follows the actions of $P_1(P_2)$ step by step through n steps. After following $P_1(P_2)$ through n steps or to termination, whichever occurs first, the program P_S signals whether or not $P_1(P_2)$ terminates by n steps, and in the termination case also signals the exact number of steps taken and what the $P_1(P_2)$ output is. In the use made of P_S in Chapter 3 to prove each of the three key nonsolvable problems, the program P_1 is always a program with one input argument and one output argument.

The program that does the task required of P_S is `execute_program`, which simulates an ideal computer program P in order to follow what happens during P's first n steps. The program P must be of type `>1<1` or `>0<1`, that is, P takes either 1 input argument or no input argument, and P returns 1 output argument. The input of `execute_program` in v_1, v_2, and v_3, respectively, is the P name, the input to P (or an empty string if P takes no input), and n, the number of P steps to be simulated. The three argument output of the program in v_1, v_2, and v_3, respectively, is a 1 or a 0 according to whether P does or does not terminate by n steps, the P output value, and the exact number of steps P takes to termination. The v_2 and v_3 output values are nonempty only if the v_1 output value is 1. If you set n to 0 instead of a positive integer, then the program is instructed to follow the execution of P to termination no matter how many steps P takes.

If you load `execute_program` into the entry form, you can do a step-by-step simulation of various programs already described that are of type `>1<1`, such as `increment_natural_integer`, `pi_d`, and `sqrt2_d`. After supplying a program name, one enters the program input, and then the number of steps to be simulated—or 0 if the simulation is to continue to P termination.

We describe in general terms the action of execute_program. The source code for this program and for its various subroutines gives a more detailed explanation. The program execute_program first calls name_to_program in order to replace the P name by the P program, and then has the P main segment and all the P subroutine segments together in v_4 as a long string. The program execute_program calls the subroutine get_named_segment to retrieve any desired segment from the P program.

The execute_program simulates the steps of a program segment Q by manipulating a certain variable string V containing in coded form all of Q's variables in serial order. The code form of a variable's string is obtained by enquoting the variable's contents, converting a quote in the string to the backslash-quote combination, and a backslash in the string to the backslash-backslash combination. Before the first step of Q is simulated, V is set to represent the initial input variables of Q, these being listed serially in enquoted form. The contents of Q's variable v_i can be retrieved from V by stepping over $i-1$ enquoted strings and retrieving the ith enquoted string, unquoting it in the process. If there are fewer enquoted strings in V than i, this indicates that variable v_i is empty. Similarly, the changed contents of a variable v_i can be recorded in V by first stepping past $i-1$ strings, adding additional enquoted empty strings if there are fewer than $i-1$ strings to step past, deleting the ith string if it is present, and finally enquoting v_i's contents as the new ith string.

The segment Q may be the main segment of P or a subroutine segment. For instance, to begin the simulation of P, execute_program obtains P's main segment by calling get_named_segment and creates an appropriate V string to represent the initial input to P. The simulation of a segment begins with the first step of the segment. After a step is simulated, the serial position of the next step to be simulated is calculated, and after first incrementing by 1 its count of executed steps, execute_program retrieves and simulates that step. Thus execute_program obtains the steps of a segment by retrieving the steps one by one from the segment string, and executes the steps by retrieving and changing appropriate variables held by V.

There are two step types that require special processing beyond the measures described, and these are the "call" and the "return" types. When execute_program encounters a call step, it enlarges the segment string it is following by putting the current V string after the beginning brace '{', and putting just before the terminal brace '}' of the segment a serial number to identify the step following the call. This enlarged string representing the steps and variable contents of the calling program is appended to a certain string R that execute_program uses when it encounters a return step. To complete its processing of a call step, execute_program retrieves the called program segment from the P program string created by name_to_program, and execute_program creates a new V string for the called segment by abbreviating appropriately the old V string, according to the number of variables the calling program supplies to the called program. After all this is done, `execute_program` can follow the steps of the called program in the way already described.

When `execute_program` encounters a return step, it retrieves the last segment appended to R, decomposes this segment back into an ordinary program segment, a V string, and a step serial number. A part of this retrieved V string is combined with a part of the current V string, according to the number of arguments the called program returns to the calling program, and the combined string becomes the new V string of the calling program. After this is done, `execute_program` resumes the simulation of the steps of the calling program at the step indicated by the serial number retrieved. If `execute_program` finds an empty R string when it attempts to simulate a return step, this indicates that the simulated program P has terminated. In this case `execute_program` ends the simulation, leaves the appropriate values in variables v_1, v_2, and v_3, and returns. The simulation also ends whenever `execute_program` finds that the number of steps simulated matches n, the number specified.

5.23 Programs That Decide Whether a Real Number Is Zero

Recall that in Section 3.4, $P^\star$ was presumed a program that could decide whether any D approximation algorithm defined the real number zero. The program $P^\star$ takes a single input argument, the name of a D approximation algorithm, and returns either 0 or 1 according to whether the algorithm does or does not define the number zero. In Section 3.4 the refuting real number was denoted by the letter z. The program that plays the role of $z(D)$ has the name `refute_d`, in accordance with our convention of identifying `real_d` program names with the two trailing symbols '`_d`'.

We will describe several different programs that play the role of $P^\star$. These programs all receive as input the name of a `real_d` program and then create another program with the name `decide` for actually carrying through the decision process and giving the $0-1$ result. Let us suppose a decision program that behaves this way has the name `decide_abc`. When we execute `decide_abc`, the program's type leads to the execution first of `decide_abc` and then, immediately afterwards, the execution of `decide`.

Decision programs are handled this way in order to obtain their output quicker. (The issue of execution speed also arose in Section 5.21 with the semifunction programs.) A decision program $P^\star$ may examine the steps of its supplied program P, and also may sample various outputs of P, that is, obtain a finite list of $P(D)$ values for $D = D_1, D_2, \ldots, D_k$. The program $P^\star$

can obtain all the steps of P by calling `name_to_program`, and $P^\star$ can obtain $P(D)$ by calling `execute_program`. But `execute_program` is slow and takes a huge number of steps in order to simulate a single P step. If $P^\star$ is a decision program that obtains $P(D)$ values by calling `execute_program`, then a more than hundred times faster $P^\star$ program is obtained by converting these `execute_program` calls to $P(D)$ calls. This requires revising a $P^\star$ program so that it takes the input P name and creates a second program to carry through the decision process, this second program obtaining $P(D)$ values by calls to P as a subroutine.

An interesting byproduct of this two-step execution of a decision program, with the decision procedure always carried out by a program named `decide`, is that the source code for `refute_d` can be written once and never needs any changes as we test various decision programs. To compose the program `refute_d`, we presume that the program `decide` is making a decision about `refute_d`, so we make `refute_d` use the procedure described in Section 3.4 to prove `decide` wrong. Thus `refute_d`, with its D input, first calls `execute_program`, supplying this subroutine three arguments. The program name `decide` is the first argument, and because `decide` is a '`>0<1`' program taking no input, the second argument is the empty string. The third argument, the n value, is D. If the first argument returned by `execute_program` is 0, signaling that `decide` does not terminate in D steps, then the D decimal interval that `refute_d` returns is `0.00` $\cdots$ `00+/-1`, an interval containing the zero point. On the other hand, if the first argument returned by `execute_program` is 1, then `refute_d` examines the second argument returned, which is `decide`'s 0–1 value. If this second argument is 1, signaling the `decide` verdict "unequal to zero," then `refute_d` behaves just as before and returns a D decimal interval containing the zero point. If this second argument is 0, signaling the `decide` verdict "equal to zero," then `refute_d` examines the third argument returned by `execute_program`, this third argument being the number of steps N taken by `decide` to come to its decision. Now `refute_d` returns the D decimal interval `0.00` $\cdots$ `010` $\cdots$ `00+/-1`, where the single 1 digit is in decimal position N. This decimal interval excludes the zero point if there is at least one 0 digit following the 1 digit, that is, if $D > N$. So whatever `decide`'s verdict is about `refute_d`, the `decide` program is wrong.

Consider first the two simple decision programs `decide0` and `decide1`. These two programs are the simplest possible. The program `decide0` always returns 0 for any `real_d` input, and the program `decide1` always returns 1 for any `real_d` input. If you load the `decide0` program and execute it, you can see its behavior. Regardless of which `real_d` program name you supply to `decide0`, the output 0 is returned, signaling "equal to zero." If any of the `real_d` names `two_d`, `one_d`, `minus_one_d`, `zero_d`, `pi_d`, or `sqrt2_d` is entered, `decide0`'s verdict is correct only for `zero_d`.

If you execute `decide0` and supply the `real_d` name `refute_d`, the `decide0` program gives its usual 0 output, but it is interesting now to execute `refute_d` to see the various D values returned by `refute_d`. We obtain the approximations shown here, the letter z denoting the real number defined by `refute_d`:

$$\begin{aligned} &z(D) \\ z(1) &= 0.0^{\pm}1 \\ z(2) &= 0.01^{\pm}1 \\ z(4) &= 0.0100^{\pm}1 \\ z(10) &= 0.0100000000^{\pm}1 \end{aligned}$$

The z number is unequal to zero, so `decide0` is definitely wrong in its conclusion about `refute_d`. The particular $z(D)$ approximations given here occur because the `decide0` conclusion about `refute_d` is made in just two steps, so when `refute_d` for $D \geq 2$ obtains the three return values 1, 0, 2 from its call of `execute_program`, it switches from providing zero approximations to providing 0.01 approximations. The constructed program `decide` is not executable because its type specification is missing. Still, one can verify that `decide` takes just 2 steps to arrive at its decision by either examining the ideal computer code of `decide` or be executing `execute_program`, entering `decide` in response to the program name request, and entering a 0 in response to the number-of-steps request.

The other simple decision program `decide1` can be tested similarly. Regardless of which `real_d` program name you supply to `decide1`, the output 1 is returned, signalling "unequal to zero." If any of the `real_d` names `two_d`, `one_d`, `m_one_d`, `zero_d`, `pi_d`, or `sqrt2_d` is entered, `decide1`'s verdict is correct for all except `zero_d`. If we enter the `real_d` name `refute_d`, we of course continue to obtain the verdict "unequal to zero." If we execute `refute_d` afterwards, we obtain the various approximations shown here:

$$\begin{aligned} &z(D) \\ z(1) &= 0.0^{\pm}1 \\ z(2) &= 0.00^{\pm}1 \\ z(4) &= 0.0000^{\pm}1 \\ z(10) &= 0.0000000000^{\pm}1 \end{aligned}$$

The `decide` program generated by `decide1`, like the `decide` program generated by `decide0`, makes its decision about `refute_d` in only two steps. However, now `refute_d` for $D \geq 2$ obtains the three return values 1, 1, 2 from its call of `execute_program`, so `refute_d` provides zero approximations for every D.

The program `decide_a` is a more sophisticated decision program. This program, when given a `real_d` name for an approximation algorithm $a(D)$, constructs a better `decide` subroutine. Its `decide` program calculates $a(1)$, and if this decimal interval contains the zero point, `decide` returns 0, otherwise it returns 1. When `decide_a` is executed with its input being any of the standard `real_d` numbers listed in the preceding paragraph, the correct decision is returned. If we execute `decide_a` and supply the input `refute_d`, we obtain the decision 0. This decision is wrong, of course, but the fact that it is wrong is harder to verify than it was for `decide0`. The program `refute_d` is chameleonlike and changes every time its name is supplied to a decision program, because the D values of `refute_d` are tied to what the newly constructed `decide` program does. When we execute `refute_d`, the decimal intervals obtained for any moderate-sized D input always include the zero point, so as far as we can tell this way, `refute_d` might be zero. The D input value we must give to `refute_d` to obtain a decimal interval excluding the zero point is quite large. To find it, we need to execute `execute_program`, respond to the request for a program name by entering `decide`, and respond to the request for an n value by entering 0 (simulate to termination). After perhaps hours of computation, the three output values returned by `execute_program` are 1, 0, and 160038. We see that 160,038 steps are needed by `decide` to reach its conclusion about `refute_d`. Why so many steps when `decide` is only obtaining $z(1)$, that is, calling `refute_d` with D set to 1? The large number of steps occurs because `refute_d`, before returning $z(D)$, calls `execute_program` with a D step simulation request, and even though `execute_program` simulates only 1 step because $D = 1$, it takes this program many steps to get prepared for the simulation. The `execute_program` calls `name_to_program` to get all the program segments, and this subroutine call alone takes many steps to assemble the program of `decide`. The previous decision programs `decide0` and `decide1` ignore whatever program name they are supplied, so the `decide` program they construct can give a decision in just two steps. The `decide_a` program requires some information about the program it is given, namely the $D = 1$ output value, and obtaining this value for `refute_d` apparently takes most of the 160,038 steps. We know now how large to set D in order to obtain from `refute_d` a decimal interval that excludes the zero point. If we execute `refute_d` and respond to the request for a D value by entering 160040, after a long wait we get a huge number of 0 digits that end with $\cdots$ `00100+/-1`. Therefore, `decide_a` is wrong as expected.

The program `decide_a` is wrong about `refute_d`, but `decide_a` is also wrong about more straightforward `real_d` programs if for $D = 1$ they return a decimal interval containing the 0 point but do not define a nonzero number. For instance, if we convert the rational 0.001 to a `real_d` number and supply the name of the constructed real_d program to decide_a, we obtain the incorrect decision "equals zero." (To carry out this experiment, execute rational_to_real_type_k and then execute real_type_k_to_real_type_d to obtain a real_d name to give decide_a.)

The next improvement over decide_a is the program decide_b. This program's decide, after obtaining a $D = 1$ approximation interval containing the zero point, the ambiguous case, attempts to resolve the ambiguity by obtaining a succession of D approximation intervals for $D = 11, 21, 31$ and 41, and only if all these intervals contain the zero point does it finally give the "equals zero" verdict. In all other cases, it correctly signals "unequal to zero." If we test this program with refute_d, we obtain the "equals zero" verdict after a lengthy computation, but as before, we can be sure that refute_d is again nonzero, with the D value needed to obtain a positive interval now much larger than the D value needed for the decide_a case.

The program decide_b can, of course, be improved further, perhaps by attempting an internal examination of the specified real_d program, but it is clear that no amount of sophistication or cleverness invested in improving the decision program can prevent its failure with refute_d.

CHAPTER 

Limits

6.1 Limit of a Sequence

Now that real numbers, sequences, and functions have been defined, we can consider limits. We apply this concept first to sequences, imitating the definition of conventional calculus.

DEFINITION 6.1: *A sequence a_n converges to a limit L,*

$$\textit{written} \lim_{n\to\infty} a_n = L$$

if there is a semifunction $N(\epsilon)$ defined for positive ϵ, such that

$$n > N(\epsilon) \quad \textit{implies} \quad |a_n - L| < \epsilon \tag{6.1}$$

The definition of a Cauchy sequence also is similar to the conventional calculus definition:

DEFINITION 6.2: *A sequence a_n is a Cauchy sequence if there is a semifunction $N_C(\epsilon)$ defined for positive ϵ, such that*

$$n, m > N_C(\epsilon) \quad \textit{implies} \quad |a_n - a_m| < \epsilon \tag{6.2}$$

Here $n, m > N_C(\epsilon)$ is an abbreviation for $n > N_C(\epsilon)$ and $m > N_C(\epsilon)$.

The next theorem is also valid in conventional calculus.

THEOREM 6.1: A sequence converges to a limit if and only if it is a Cauchy sequence.

Assume a_n is a sequence converging to a limit L. If $n, m > N(\epsilon/2)$, we have

$$|a_n - a_m| = |a_n - L + L - a_m| \leq |a_n - L| + |L - a_m| < \frac{\epsilon}{2} + \frac{\epsilon}{2} = \epsilon$$

We can take $N_C(\epsilon)$ equal to $N(\epsilon/2)$, and a_n is a Cauchy sequence.

Next assume a_n is a Cauchy sequence. We show that a_n has a limit L with approximation algorithm $L(E)$. Given an argument E, the computation for $L(E)$ consists of choosing a positive integer n_E larger than $N_C(E/2)$, and then using the a_n sequence program to obtain an interval $a_{n_E}(E/2) = m \pm e$. Then $L(E)$ is taken to be the widened interval $m \pm (e + E/2)$, both interval endpoints having been moved outward by the amount $E/2$. To be certain a real number has been defined, we must show that for any E_1 and E_2, the intervals $L(E_1)$ and $L(E_2)$ intersect. As shown in Section 2.2, the real number $a_{n_{E_1}}$ lies in the interval $a_{n_{E_1}}(E_1/2)$ and the real number $a_{n_{E_2}}$ lies in the interval $a_{n_{E_2}}(E_2/2)$, while the distance separating $a_{n_{E_1}}$ and $a_{n_{E_2}}$ is restricted by the Cauchy sequence inequality

$$|a_{n_{E_1}} - a_{n_{E_2}}| < \max(E_1/2, E_2/2)$$

Therefore, the extended intervals $L(E_1)$ and $L(E_2)$ must intersect. Note that the real number L and the real number a_{n_E} both lie in the interval $L(E)$ of length at most $2E$, so $|a_{n_E} - L| \leq 2E$.

To show that L is the limit of the sequence a_n, we note that if $n > N_C(\epsilon/2)$, and E_0 is a rational number chosen $< \epsilon/4$, we have

$$\begin{aligned} |a_n - L| &= |a_n - a_{n_{E_0}} + a_{n_{E_0}} - L| \\ &\leq |a_n - a_{n_{E_0}}| + |a_{n_{E_0}} - L| \\ &< \frac{\epsilon}{2} + 2E_0 < \frac{\epsilon}{2} + \frac{\epsilon}{2} = \epsilon \end{aligned}$$

Therefore the semifunction $N(\epsilon)$ may be taken equal to $N_C(\epsilon/2)$.

In conventional calculus, many properties of sequence limits are proved, and as such proofs are constructive, they go over without difficulty in computable calculus. We list here, without repeating the conventional calculus proofs, a number of properties of limits:

(i) If $\lim_{n\to\infty} a_n = A$ and $\lim_{n\to\infty} b_n = B$, then $\begin{cases} \lim\limits_{x\to\infty} (a_n + b_n) = A + B, \\ \lim\limits_{x\to\infty} (a_n - b_n) = A - B, \\ \lim\limits_{x\to\infty} (a_n b_n) = AB. \end{cases}$

(ii) If $\lim_{n\to\infty} a_n = A$ and $\lim_{n\to\infty} b_n = B$, then $a_n \leq b_n$ for $n > n_0$ implies $A \leq B$.

(iii) If $\lim_{n\to\infty} a_n = A = \lim_{n\to\infty} b_n$, and c_n is a sequence satisfying $a_n \leq c_n \leq b_n$ for $n > n_0$, then $\lim_{n\to\infty} c_n = A$.

6.2 Monotone Sequences

DEFINITION 6.3: *A sequence a_n is* monotone increasing *if $a_{n+1} \geq a_n$ for all n, and is* strictly monotone increasing *if $a_{n+1} > a_n$ for all n. A* monotone decreasing *and a* strictly monotone decreasing *sequence are defined analogously. A sequence a_n is* bounded *if there is a number M such that $|a_n| \leq M$ for all n.*

We often deal with pairs of monotone sequences, and the next theorem, valid in both computable and conventional calculus, is relevant for this case.

THEOREM 6.2: If the sequence a_n is monotone increasing and the sequence b_n is monotone decreasing, and $\lim_{n\to\infty} (b_n - a_n) = 0$, then both sequences a_n and b_n converge to the same limit.

We must have $a_n \leq b_n$ for all n. For if for some n_0 we had $a_{n_0} > b_{n_0}$, then because b_n is monotone decreasing and a_n is monotone increasing, for $n > n_0$ we would have $a_n - b_n \geq a_{n_0} - b_{n_0} > 0$, implying $\lim_{n\to\infty} (a_n - b_n) > 0$. However, $\lim_{n\to\infty} (a_n - b_n) = \lim_{n\to\infty} -1 \cdot (b_n - a_n) = -1 \cdot 0 = 0$, so we obtain a contradiction.

For the given limit, $\lim_{n\to\infty} (b_n - a_n) = 0$, there is a semifunction $N(\epsilon)$ such that $n > N(\epsilon)$ implies $|b_n - a_n| = |(b_n - a_n) - 0| < \epsilon$. The semifunction $N(\epsilon)$ can also serve as the semifunction $N_C(\epsilon)$ needed to show a_n is a Cauchy sequence, because if $n, m > N(\epsilon)$, then

$$\begin{aligned} |a_n - a_m| &= a_{\max(n,m)} - a_{\min(n,m)} \leq b_{\max(n,m)} - a_{\min(n,m)} \\ &\leq b_{\min(n,m)} - a_{\min(n,m)} = |b_{\min(n,m)} - a_{\min(n,m)}| < \epsilon \end{aligned}$$

Hence a_n is Cauchy and converges to a limit L. The sequence b_n converges to the same limit because

$$\lim_{n\to\infty} b_n = \lim_{n\to\infty} [a_n + (b_n - a_n)] = L + 0 = L$$

A standard result in conventional calculus is that a bounded, monotone sequence converges to a limit. Let us consider the usual proof of this result. Let a_n be a monotone increasing sequence, and let M be a positive number such that $|a_n| \leq M$ for all n. Two other sequences l_N and u_N are now defined with $u_N - l_N = 2M/2^{N-1}$ for all N, such that l_N is monotone increasing, u_N is monotone decreasing, and for every N there are infinitely many terms a_n satisfying the relations $l_N \leq a_n \leq u_N$. We take l_1 equal to $-M$ and take u_1 equal to M. All the requirements specified are true for the case $N = 1$. Next, assuming that l_N and u_N are defined with all requirements true for this N, we define l_{N+1} and u_{N+1} recursively in terms of l_N and u_N as follows: Let m_N be the midpoint $(l_N + u_N)/2$ of the interval $[l_N, u_N]$. There are infinitely many terms a_n in the interval $[l_N, u_N]$. Either there are infinitely many terms a_n in the interval $[l_N, m_N]$, or there are infinitely many terms in the interval $(m_N, u_N]$. Only one of these intervals can hold infinitely many terms. This is because a_n is monotone increasing, so any term at all in $(m_N, u_N]$ implies only finitely many terms in $[l_N, m_N]$. If there are infinitely many terms in the interval $(m_N, u_N]$, we take l_{N+1} equal to m_N and take u_{N+1} equal to u_N. If there are infinitely many terms in the interval $[l_N, m_N]$, we take l_{N+1} equal to l_N and take u_{N+1} equal to m_N. All requirements now are true for the $N + 1$ case.

The equation $u_N - l_N = 2M/2^{N-1}$ implies $\lim_{N\to\infty} (u_N - l_N) = 0$, and then using the theorem just proved, we obtain $\lim_{N\to\infty} u_N = \lim_{N\to\infty} l_N = L$. Because there are an infinite number of terms a_n satisfying the inequalities $l_N \leq a_n \leq u_N$, and the sequence a_n is monotone increasing, there are no terms a_n greater than u_N, so we must conclude that the sequence a_n also converges to L.

This proof does not carry over into computable calculus because even though it is logically clear that there are infinitely many a_n terms in either $[l_N, m_N]$ or $(m_N, u_N]$, we have not shown how to decide "by finite means" which interval it is that contains infinitely many terms, and we need an explicit procedure if l_N and u_N are to be computable sequences. It appears doubtful that the decision as to which of the two intervals contains infinitely many a_n terms can be resolved by an ideal computer in a finite number of steps. Indeed, Specker's theorem in the next section makes it clear that every constructive method of making this decision is certain to fail in some cases.

It is illuminating at this point to consider a somewhat different question: Are there *finite* sequences l_N and u_N, defined for N equal to the values 1, 2, 3, up to some huge number H, that have the correct values assigned according to the infinite a_n terms selection system just described? This time we would have to agree that such finite sequences exist, simply because one could construct all possible sequence pairs taking all possible varieties of the choices allowed, and one of these finite sequence pairs must have its values correctly assigned. That is, if we make up 2^{H-1} different pairs of finite sequences l_N and u_N according to this number of ways of arbitrarily making choices between the leftmost interval $[l_N, m_N]$ and the rightmost interval

$(m_N, u_N]$ for every N between 2 and H, one of our constructions will surely be correct. Accordingly, there is no difficulty agreeing that there is a finite sequence pair that is correctly formed up to N equal H, where H is some arbitrarily large number. Consequently, it is acceptable to consider further properties that a finite sequence pair l_N and u_N would have, the sequence pair formed according to the plan given. That is, we could consider making further deductions about the properties of the finite sequence terms l_N and u_N.

However, a finite sequence pair does not serve to define a common limit. For this, a sequence pair defined for all N is needed, and to obtain this, we need to specify how the sequence pair can be programmed on a ideal computer.

This digression does show, however, that in computable calculus, when an argument requires considering a *finite* number of possibilities, any difficulty in choosing among the possibilities may safely be ignored. This is the *principle of finite choice*. This situation arises often in proving inequalities. For example, we may need to consider the case $x = 0$ and the case $x \neq 0$ separately. The difficulty in computably distinguishing between the two cases is of no consequence. Whenever there are an infinite number of possibilities, the situation changes, for if we allow the previous reasoning, we escape the ideal computer constructivity.

6.3 The Specker Theorem

In 1949 the Swiss mathematician Ernst Specker [34] proved the following theorem.

THEOREM 6.3: There is a strictly monotone increasing and bounded sequence b_n that does not converge to a limit.

Before defining b_n, we first give a procedure for determining whether a real number a with approximation algorithm $a(D)$ is unequal to zero. The procedure is certain to be eventually successful if the number a is nonzero. In succession, we obtain the intervals $a(1), a(2), a(3), \ldots$, stopping as soon as we find for some D that $a(D) \dot{\neq} 0$, for then a is surely nonzero. Of course if it happens that a equals zero, then our procedure does not terminate, and we compute forever.

In Section 4.2 we showed that programs P of our ideal computer can be associated with the positive integers, each program P being associated with an integer $<P>$, obtained by viewing P as an integer in the S-ary number system, where S is the number of symbols needed to specify ideal computer programs. Conversely, each positive integer i can be associated with a program P_i, the program P_i found by converting i into the S-ary system, and then interpreting the digits of i as programming symbols.

Now we can define b_n. Each term b_n is an n-digit decimal number of the form

$$b_n = 0.d_1^{(n)} d_2^{(n)} \ldots d_n^{(n)}$$

Here a digit $d_i^{(n)}$ is either 2 or 7, according to the following plan. It is presumed that P_i is the type D approximation algorithm of some real number, and the procedure just described for detecting zero inequality is applied to P_i, the computation being followed through to n steps. If the computation to this number of steps has revealed that the number defined by P_i is nonzero, then $d_i^{(n)} = 7$, and if it has not, then $d_i^{(n)} = 2$. The presumption that P_i is a type D approximation algorithm of course may be incorrect, but this does not matter. The digit $d_i^{(n)}$ equals 7 only if it is found for a certain D, say D_0, that $P_i(D_0)$ defines an interval with rational endpoints not containing the zero point, this result detected after following through at most n steps of the computations for $P_i(1)$, $P_i(2)$, etc. In any other case, including the case of a defective program P_i, the digit $d_i^{(n)}$ equals 2.

Thus the ith digit $d_i^{(n)}$ for low values of n will equal 2, and perhaps shifts to the value 7, and stays at 7, only for $n \geq N$, where N is the exact number of steps needed to calculate $P_i(1), P_i(2), \ldots, P_i(D_0)$, where D_0 is the smallest input integer resulting in an output interval excluding the zero point. If this shift to 7 never occurs, then $d_i^{(n)}$ equals 2 for every n for which $d_i^{(n)}$ is defined, that is, for $n \geq i$. Consequently a limit L for the sequence b_n must be such that its decimal digits are only 2 or 7.

As the decimal digits of b_n are never smaller than the decimal digits of b_{n-1}, and b_n has the nonzero decimal digit $d_n^{(n)}$ that b_{n-1} does not have, the sequence b_n is strictly monotone increasing. The sequence is also bounded because for all n we have

$$0 < b_n < 0.777 \cdots 77 \cdots = \frac{7}{9}$$

The sequence b_n cannot converge to a limit L with approximation algorithm $L(D)$, because we would have a way of deciding whether any real number was zero, contradicting Nonsolvable Problem 3.1. Given a real number a with approximation algorithm $a(D)$ computed by a program P, we find the integer $i = \lt P \gt$ and then evaluate $L(D)$ with D taken equal to $i + 1$ to determine whether the ith decimal digit of the approximant is 2, indicating $a = 0$, or is 7, indicating $a \neq 0$.

6.4 Consequences of Specker's Theorem

The failure in computable calculus of the theorem that every monotone bounded sequence converges to a limit does not prevent obtaining the usual calculus results concerning infinite series. Instead of trying to show for an infinite series that the sequence of its partial sums has a limit because this sequence is monotone and bounded, one shows that the sequence has a limit because it is a Cauchy sequence. However, there are other, more fundamental, repercussions from the Specker result. Consider least upper bounds or greatest lower bounds:

DEFINITION 6.4: *A number M is an* upper bound *of a sequence a_n if $a_n \leq M$ for all n. If a_n is a sequence with an upper bound, the* least upper bound *of a_n is an upper bound M such that $M \leq M'$ where M' is any upper bound for a_n. A* lower bound *and* greatest lower bound *of a sequence a_n are defined analogously.*

In conventional calculus, a sequence with an upper bound always has a least upper bound. In computable calculus, this is not the case.

THEOREM 6.4: A sequence exists with an upper bound but without a least upper bound.

The sequence of this theorem is the Specker theorem sequence b_n with upper bound $\frac{7}{9}$. If M, with approximation algorithm $M(D)$, is a least upper bound for the strictly monotone increasing sequence b_n, then M must be the limit of the sequence, with the same contradiction as before.

6.5 Limit of a Function

In what follows we define the limit of a function in almost the same way this concept is defined in conventional calculus. This definition is labeled tentative because an adjustment of the definition proves advisable.

TENTATIVE DEFINITION 6.5: *A function $f(x)$ has a limit L as x approaches x_0,*

$$\text{written } \lim_{x \to x_0} f(x) = L$$

if $f(x)$ is defined on an interval (a, x_0) and on an interval (x_0, b), and there is a positive-valued semifunction $\delta(\epsilon)$ of the positive argument ϵ, such that for any x

for which $f(x)$ *is defined,*

$$0 < |x - x_0| < \delta(\epsilon) \quad \textit{implies} \quad |f(x) - L| < \epsilon$$

The function $f(x)$ *is* continuous *at* x_0 *if* $f(x)$ *is defined at* $x = x_0$ *and* L *equals* $f(x_0)$*, that is, if*

$$\lim_{x \to x_0} f(x) = f(x_0)$$

The semifunction $\delta(\epsilon)$ *is called the* limit modulus.

Our definition of a function $f(x)$ specifies a program whose input is an $x(K)$ approximation algorithm, and whose output is a $y(K)$ approximation algorithm. Let $\widehat{f}(x)$ now be one of the functions considered in Section 4.3, such as a polynomial, rational function, or algebraic function. For these functions we described in Section 4.3 how the algorithm $y(K)$ defining a function value can be generated using interval arithmetic operations that start with the interval $x(K)$. Each of these functions is continuous at any argument x_0 interior to the functions domain interval I, and we now describe a general method of forming the required limit modulus $\delta(\epsilon)$ required by the preceding definition.

Recall that a $y(E)$ approximation algorithm is obtained from $y(K)$ by computing in succession $y(1), y(2), y(3), \ldots$, and returning the first y interval with an error bound not exceeding E. If x_0 is a domain interior point and y_0 is $\widehat{f}(x_0)$, then for any positive rational E, there is a corresponding K parameter value K_E, such that the interval $y_0(K_E)$ has an error bound not exceeding E. We know that $y_0(K_E)$ is generated using $x_0(K_E)$. Let us agree for the time being that this implies that every real number in the interval $x_0(K_E)$ is mapped by $\widehat{f}(x)$ into the interval $y_0(K_E)$. (We address this point later.) This means that all real numbers in the interval $x_0(K_E)$ are mapped by $\widehat{f}(x)$ into an interval of length not greater than $2E$ (see Fig. 6.1).

The first step in the $\delta(\epsilon)$ construction is to replace the approximation algorithm $x_0(K)$ supplied to the $\widehat{f}(x)$ program by the more convenient algorithm $\widehat{x}_0(K)$, having the same approximants as $x_0(K)$ but with an error bound twice as large. Doing this makes it certain the number x_0 is within the middle half of the interval $\widehat{x}_0(K)$, that is, the part of $\widehat{x}_0(K)$ that equals $x_0(K)$. The value, K_E, taken by K when the interval $y_0(K)$ first has an error bound not exceeding E, gives a value for $\delta(2E)$, namely $e_{K_E}/2$, where e_{K_E} is the error bound of $\widehat{x}_0(K_E)$. This is because: (1) The number y_0 is somewhere in the interval $y_0(K_E)$, of length not larger than $2E$; (2) all real numbers x in the interval $\widehat{x}_0(K_E)$ are mapped by $\widehat{f}(x)$ into $y_0(K_E)$; and (3) the number x_0 is somewhere in the middle half of $\widehat{x}_0(K_E)$, so the $\widehat{f}(x)$ image of any x within $e_{K_E}/2$ of x_0 is within $2E$ of $f(x_0)$.

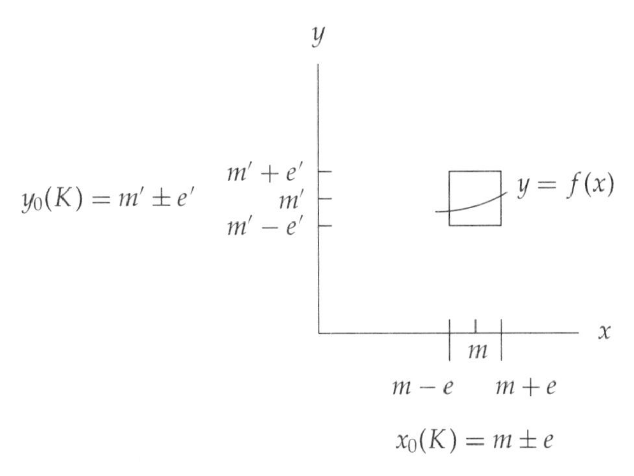

Figure 6.1. The function $f(x)$ near $x = x_0$.

The semifunction $\delta(\epsilon)$ now can be taken equal to $e_{K_E}/2$ where $2E$ is a positive rational number chosen less than ϵ by some fixed procedure. In a sense we mislead the $\widehat{f}(x)$ program purposely, supplying it overly wide x_0 intervals through $\widehat{x}_0(K)$, knowing the function value approximation interval returned will use these intervals in its computations for $y_0(K)$, and thus help us to obtain $\delta(\epsilon)$. This method of computing $\delta(\epsilon)$ for any argument x_0 interior to the domain interval I is successful only because we know the computation method used by these $\widehat{f}(x)$ programs is the interval procedure described in Chapter 4.

Notice here the convenience of having $\delta(\epsilon)$ be a semifunction instead of a function. If a function $\delta(\epsilon)$ were required, then for its construction from the values e_{K_E}, we would have to employ some more elaborate system, and perhaps use linear interpolation as sketched in Fig. 6.2.

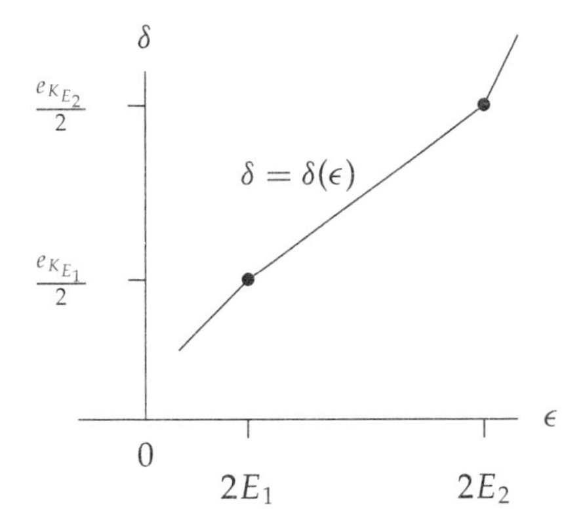

Figure 6.2. A function $\delta(\epsilon)$ obtained by linear interpolation.

Now we address the question of whether every real number in the interval $x_0(K_E)$ is mapped by $\widehat{f}(x)$ into the interval $y_0(K_E)$. It is easy to see that every rational number p/q in $x_0(K_E)$ is mapped into $y_0(K_E)$, because with $x_0(K_E)$ replaced by $(p/q) \pm 0$, the interval computation procedure now starts with a point interval contained within $x_0(K_E)$, and so is certain to yield a result interval within $y_0(K_E)$. We now can use the reasoning employed in Section 4.6 and conclude that every real number in $x_0(K_E)$ is mapped into $y_0(K_E)$.

The next theorem is fundamental.

THEOREM 6.5: (Ceitin [8].) A real function $f(x)$ defined in a finite or infinite interval I is continuous at every interior point x_0 of I.

Let $x_0(K)$ be the approximation algorithm for x_0, and let $y_0(K)$ be the approximation algorithm returned by $f(x)$. Given any E, with $y_0(K)$ we are able to find a K_E, such that $y_0(K_E) = m \pm e$ with $e \leq E$. We expect the approximation algorithm $y_0(K)$ to depend on the algorithm $x_0(K)$. If we could somehow discover an interval $x_0(K'_E)$ which is mapped into $y_0(K_E)$, then we could construct the semifunction $\delta(\epsilon)$ by the method just described for the example functions. For an example function $\widehat{f}(x)$, the integer K'_E is equal to K_E. However, we do not know how an arbitrary $f(x)$ program will compose $y_0(K)$ so we cannot presume anything about K'_E. Nevertheless, as Ceitin showed, because $f(x)$ is a function and not a semifunction, the $f(x)$ program can be forced to divulge information that is equivalent to supplying us with an interval $x(K'_E)$ for any E.

Given a particular positive rational E_0, to obtain K'_{E_0} for an arbitrary function $f(x)$ at an interior point x_0 defined by an arbitrary approximation algorithm $x_0(K)$, we define a certain function $z(K)$, which may or may not be an approximation algorithm, that behaves as follows: For any K, it uses the subroutine P_S described in Chapter 3, to follow the computations made by the $f(x)$ program and subsequent computations for $y(1), y(2), \ldots, y(K_{E_0})$ through a total of K steps to determine whether by this number of steps, a value for K_{E_0} is found with $y(K_{E_0}) = m \pm e$, and with $e \leq E_0$. If termination for all these computations does not occur by K steps, then $z(K)$ equals $x_0(K)$. If termination occurs after precisely N steps yielding the $f(z)$ approximation interval $m \pm e$, with $e \leq E_0$, then systematically, for every rational r in $x_0(N)$, the z program computes, using its own stored copy of the $f(x)$ program, a sequence of ever more accurate approximations to $f(r)$, attempting to find a rational r in $x_0(N)$ such that $f(r)$ does not lie in $m \pm e$. If such a rational is not found, the computation is endless and $z(K)$ is undefined for $K \geq N$. If r_z is the first rational found with the required properties, then $z(K)$ for $K \geq N$ equals $r_z(K)$, that is, $r_z \pm 0$.

Let us make more explicit the computation that $z(K)$ does in the termination case. There is no difficulty obtaining all the rational numbers in $x_0(N)$ as a sequence r'_n, with no rational term repeated. To find an nth rational in $x_0(N)$, we use the sequence r_n of all rational numbers, defined in Section 4.1, test these sequence terms one by one for inclusion in $x_0(N)$, and discard those outside this interval, until we arrive at the nth rational in $x_0(N)$, and this rational becomes r'_n. The ever more accurate values for $f(r)$ can be obtained by finding first an $E = 10^{-1}$ interval for $f(r'_1)$, then $E = 10^{-2}$ intervals for both $f(r'_1)$ and $f(r'_2)$, and so forth.

Now suppose we supply the program $z(K)$ to $f(x)$ and attempt to obtain an interval $m \pm e$ with $e \leq E_0$ from the algorithm $y(K)$ returned by the $f(x)$ program. The program $z(K)$ is not necessarily a real number approximation algorithm. Will $f(x)$ return an approximation algorithm $y(K)$ that yields such an interval when evaluated at progressively larger K? Such an interval must be obtainable, because otherwise this would mean $z(K)$ is equal to $x_0(K)$, yet either an approximation algorithm was not supplied by $f(x)$, or this algorithm did not have the required properties. And if the interval $m \pm e$ is determined in N steps, every rational in $x_0(N)$ is mapped into $m \pm e$, otherwise we have another contradiction, namely that $z(K)$ is defined for all K, being an approximation program for some rational r_z in $x_0(N)$, for which $f(x)$ has returned an approximation algorithm $y(K)$ defining a real number unequal to the real number defined by $f(x)$ for $z(K)$. And if every rational in $x_0(N)$ is mapped into $m \pm e$, then every real number in $x_0(N)$ is likewise mapped into $m \pm e$, by the reasoning employed in Section 4.6.

Through the program $z(K)$ we obtain K'_{E_0} as the integer N, and we can obtain any integer value K'_E we need in a similar fashion. The semifunction $\delta(\epsilon)$ required by Definition 6.5 now can be obtained by using the simple construction plan described previously. This completes the proof of the theorem.

Theorem 6.5 makes it advisable to change our tentative definition of limit for two reasons. First, taking an $f(x)$ limit as x approaches a point x_0 interior to the domain interval I of $f(x)$ is of little interest because our functions are continuous at such points, and the limit obtained is always $f(x_0)$. The tentative definition has the line shown here:

$$0 < |x - x_0| < \delta(\epsilon) \quad \text{implies} \quad |f(x) - L| < \epsilon$$

The beginning requirement $0 < |x - x_0|$ is present in order to obtain limits for functions that are not continuous at x_0. However, such functions are not possible in computable calculus, and the line shown in the preceding can be simplified to

$$|x - x_0| < \delta(\epsilon) \quad \text{implies} \quad |f(x) - L| < \epsilon$$

The chief use of the limit definition in computable calculus is to obtain limits where x_0 is a boundary point of the $f(x)$ domain, but x_0 is not part of the domain. Perhaps the domain consists of a pair of intervals with the point x_0 separating the two intervals but not being a member of either interval.

The second difficulty with the limit definition occurs when the definition is generalized to apply to functions $f(x, y)$ so limits of the form $\lim_{(x,y)\to(x_0,y_0)} f(x, y)$ can be used. Here the natural generalization of the requirement that $f(x)$ be defined on an interval (a, x_0) and on an interval (x_0, b) is the requirement that $f(x, y)$ be defined in a "neighborhood" of (x_0, y_0) except at the point (x_0, y_0) itself, that is, on some punctured disk $0 < \sqrt{(x - x_0)^2 + (y - y_0)^2} < \delta$. However, functions of two variables rarely have domains that fail to include such isolated points (x_0, y_0). If they fail to be defined at a point (x_0, y_0), it is more likely they fail to be defined at a connected set of points containing (x_0, y_0). The limit definition needs to have the condition that $f(x)$ be defined to the left and to the right of x_0 relaxed, so that its extension to functions with more than one variable is more convenient.

Our revised definition of limit is

FINAL DEFINITION 6.5: *A function $f(x)$ has a limit L as x approaches x_0,*

$$\textit{written } \lim_{x\to x_0} f(x) = L$$

if the domain of $f(x)$ includes a finite interval I that either contains x_0 or has x_0 as a boundary point, and there is a positive-valued semifunction $\delta(\epsilon)$ of the positive argument ϵ, such that for any x for which $f(x)$ is defined,

$$|x - x_0| < \delta(\epsilon) \quad \textit{implies} \quad |f(x) - L| < \epsilon$$

The function $f(x)$ is continuous *at x_0 if $f(x)$ is defined at $x = x_0$ and L equals $f(x_0)$, that is, if*

$$\lim_{x\to x_0} f(x) = f(x_0)$$

Now for a limit to exist, $f(x)$ can be defined just in (a, x_0), or can be defined just in (x_0, b). These domains provide sufficient neighboring points to x_0 for the limit concept to make sense. However, note that although a minimum domain I is required for a limit to exist, the requirement

$$|x - x_0| < \delta(\epsilon) \quad \text{implies} \quad |f(x) - L| < \epsilon$$

must hold for all points x where $f(x)$ is defined, not just for the points in the minimum domain. The foregoing definition does not exclude the case where $f(x)$ is defined in both (a, x_0) and (x_0, b), but no longer is such a domain required. The case where $f(x)$ is defined in an interval (a, b) containing x_0 is uninteresting, because then Theorem 6.5 implies the limit is $f(x_0)$.

We obtain right- and left-hand limits this way:

DEFINITION 6.6: *A function $f(x)$ has a limit L as x approaches x_0 from the right,*

$$\textit{written} \lim_{x \to x_0^+} f(x) = L$$

if the domain of $f(x)$ has a subdomain of the form (x_0, b), and with $f(x)$ imagined restricted to this subdomain, we find $\lim_{x \to x_0} f(x) = L$. Similarly, $f(x)$ has a limit L as x approaches x_0 from the left,

$$\textit{written} \lim_{x \to x_0^-} f(x) = L$$

if the domain of $f(x)$ has a subdomain of the form (a, x_0), and with $f(x)$ imagined restricted to this subdomain, we find $\lim_{x \to x_0} f(x) = L$.

As an example, consider the function sgn$x$ of computable calculus, which is defined as $x/|x|$ in the intervals $(-\infty, 0)$ and $(0, \infty)$. Here $\lim_{x \to 0} \operatorname{sgn} x$ does not exist, but according to our definition, appropriate subdomains exist for obtaining $\lim_{x \to 0^+} \operatorname{sgn} x = 1$ and $\lim_{x \to 0^-} \operatorname{sgn} x = -1$.

With our final definition of limit, we can restate Ceitin's theorem in a more general form.

THEOREM 6.6: A function $f(x)$ defined on a finite or infinite interval I is continuous at every point of I.

Thus if $f(x)$ is defined just on $[a, b]$, the theorem implies $\lim_{x \to a^+} f(x) = f(a)$ and $\lim_{x \to b^-} f(x) = f(b)$. The relaxed form of our limit definition makes it unnecessary to state as special cases the conditions that hold at the endpoints a and b.

6.6 Using Limits to Extend a Function's Domain

Suppose the function $f(x)$ is defined on an interval I. In computable calculus, $\lim_{x \to c} f(x)$ is easy to find if the point c lies in I. This limit is always $f(c)$. Limit expressions are not this simple if the point c is a boundary point of I and c does not lie in I, for then the limit must be found by some other means than function evaluation. The interval I has c as an endpoint and I is "open" at this endpoint. That is, at c parenthesis notation is used instead of bracket notation. When $\lim_{x\to c} f(x)$ exists, the function $f(x)$ defined on I can be extended to become a function defined on an interval which includes c, that is, the interval I gets "closed" at c, the parenthesis being changed to a bracket.

Similarly, if the interval I has c as an interior point and the function $f(x)$ is defined on I with the single exception of the point c, then if $\lim_{x\to c} f(x)$ exists, the function can be extended to become a function defined everywhere on I. Our relaxed definition of limit allows us to state this result in a general way. With the conventional definition, we would have to enumerate the various cases, using right- and left-hand limits.

THEOREM 6.7: Let I be an interval that contains the point c as an interior point or as a boundary point. If the function $f(x)$ is defined on I except for the point c, and $\lim_{x\to c} f(x) = L$, then $f(x)$ can be extended to become a function defined on I.

We consider only the case where $f(x)$ is defined on $[a, c)$ and $(c, b]$. Any other case can be treated similarly. For the case we consider, the extended function is

$$f_1(x) = \begin{cases} f(x) & \text{if } x \text{ is in } [a, c) \text{ or } (c, b] \\ L & \text{if } x \text{ is } c \end{cases}$$

The approximation algorithm $y(K)$, returned by the f_1 program for the input algorithm $x(K)$, uses the f program, the $x(K)$ program, the program $c(K)$ defining c, the program $L(K)$ defining L, and the $\delta(\epsilon)$ program. For any K, the rational $E = 10^{-K}$ is formed, with the goal being to return an approximation interval with error bound not exceeding E. The numbers 0 and $d = \delta(E/2)$ are distinct, so according to Solvable Problem 3.10, by forming more and more accurate approximations to $|x - c|$ and d, we can eventually determine a true inequality from the following choices: $|x - c| < 0$, $|x - c| > 0$, $|x - c| < d$, $|x - c| > d$. The first possibility cannot occur because $|x - c| \geq 0$, and if we find the second or last possibility occurs, then it is certain that $x \neq c$, and we can obtain our E approximation by using the $f(x)$ program and evaluating the returned approximation algorithm for progressively larger K. In the remaining case $|x - c| < d$, we obtain our E interval as $L(E/2)$ except that the obtained error bound is increased by the amount $E/2$ to reflect the inequality $|x - c| < d = \delta(E/2)$.

There are many applications of this result. The function $g(x)$ defined by the equation

$$g(x) = \begin{cases} x \sin(\frac{1}{x}) & \text{if } x \neq 0 \\ 0 & \text{if } x = 0 \end{cases}$$

is a computable function defined in $(-\infty, \infty)$, because for the function $x \sin(1/x)$ defined everywhere except at $x = 0$, we have $\lim_{x\to 0} x \sin(1/x) = 0$. As another example, the function $\text{sgn}\ x = x/|x|$ is not defined at $x = 0$, but nevertheless the identity $x\ \text{sgn}\ x = |x|$ may be considered valid for all x, including the case $x = 0$. This is because for the function $x\ \text{sgn}\ x$, we have $\lim_{x\to 0} x\ \text{sgn}\ x = 0$, so $x\ \text{sgn}\ x$ can be extended to the function

$$h(x) = \begin{cases} x\ \text{sgn}\ x & \text{if } x \neq 0 \\ 0 & \text{if } x = 0 \end{cases}$$

The extended function is a computable function defined for all x. For $x \neq 0$ we have $h(x) = x\ \text{sgn}\ x = x(x/|x|) = x^2/|x| = |x|^2/|x| = |x|$, and for $x = 0$ we have $h(x) = 0 = |x|$. Thus, in all cases the extended function equals $|x|$. In similar fashion the identity $|x|\ \text{sgn}\ x = x$ is valid for all x.

6.7 A Standard Result for Functions

In many elementary calculus texts, at some point there appears the statement that if a function $f(x)$ is continuous in $[a, b]$, this implies that for every number c between $f(a)$ and $f(b)$ there exists at least one number x_0 in (a, b) such that $f(x_0) = c$. The next theorem establishes a similar result for computable functions $f(x)$ defined in $[a, b]$. Computable functions do not have to be qualified as continuous, because they are necessarily continuous at any point where they are defined. Immediately following the theorem, we show that it is a nonsolvable problem to determine x_0 to k or $k+1$ correct decimals. Rather than say "a number x_0 exists" or "a number x_0 can be found," when x_0 sometimes cannot be determined accurately by finite means, it is less confusing to say "the number x_0 cannot fail to exist."

THEOREM 6.8: Suppose the function $f(x)$ is defined on $[a, b]$ with $f(a) \neq f(b)$, and c is a number between $f(a)$ and $f(b)$. In the interval (a, b), a number x_0 such that $f(x_0) = c$ cannot fail to exist.

For clarity, we assume $f(a) < c < f(b)$. The other case $f(a) > c > f(b)$ would be treated analogously. To prove the theorem, we assume no number x_0 in (a, b) exists such that $f(x_0)$ equals c, and then show that this assumption leads to a contradiction. Because of our assumption, we can define two monotone sequences, a_n and b_n, such that $f(a_n) < c < f(b_n)$, and such that $b_n - a_n = (b - a)/2^{n-1}$. We take $a_1 = a$ and $b_1 = b$, so the required conditions are true for $n = 1$. If a_n and b_n are defined and satisfy all conditions, then the terms a_{n+1} and b_{n+1} are obtained this way: Take $m_n = (a_n + b_n)/2$; the numbers c and $f(m_n)$ may be computed arbitrarily accurately. Because the two numbers are distinct, as we compute more and more accurate intervals for them, eventually we must encounter approximation intervals that do not intersect, and then we also determine which of the two intervals is to the right of the other. If we find $f(m_n) > c$, then we take a_{n+1} equal to a_n and take b_{n+1} equal to m_n. Otherwise we find $f(m_n) < c$, and then we take a_{n+1} equal to m_n and take b_{n+1} equal to b_n. Because for all n we have $b_n - a_n = (b-a)/2^{n-1}$, by Theorem 6.2 the two monotone sequences are Cauchy and have a common limit L, contained in $[a, b]$. The function $f(x)$ is continuous at $x = L$, so the sequences $f(a_n)$ and $f(b_n)$ both converge to $f(L)$. For all n we have $f(a_n) < c < f(b_n)$, so we obtain

$$\lim_{n\to\infty} f(a_n) \leq c \leq \lim_{n\to\infty} f(b_n)$$

or

$$f(L) \leq c \leq f(L)$$

This implies $f(L) = c$, so the point L is in (a, b). Therefore, x_0 can be taken equal to L, and the assumption that x_0 does not exist is wrong.

NONSOLVABLE PROBLEM 6.1: Given a function $f(x)$ defined on an interval $[a, b]$ such that $f(a) \neq f(b)$, and given a number c between $f(a)$ and $f(b)$, find to k or $k+1$ correct decimals an argument x_0 in (a, b) such that $f(x_0) = c$.

Suppose the search interval is $[-2, 2]$, the number c is zero, and $f(x)$ equals $f_a(x)$, diagrammed in Fig. 6.3 and defined by the equation

$$f_a(x) = \max(\min(a, x+1), x-1)$$

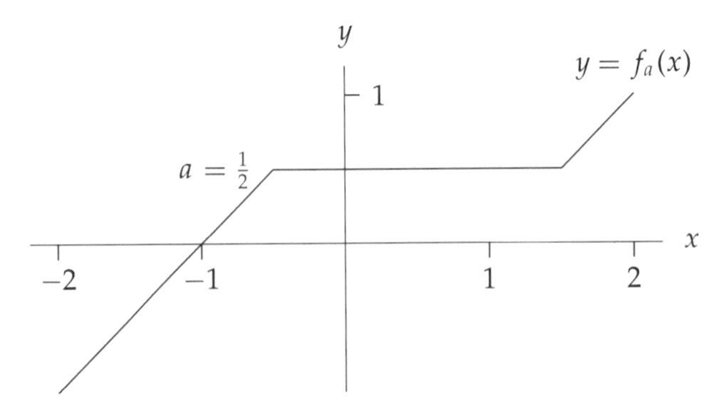

Figure 6.3. The function $f_a(x) = \max(\min(a, x+1), x-1)$ when $a = \frac{1}{2}$.

This function has an interval of width 2 on which it equals a, and is linear with slope 1 elsewhere. For any value of the parameter a, we have $f_a(2)$ positive and $f_a(-2)$ negative. If $a > 0$, then $f_a(x)$ is zero at $x = -1$; if $a < 0$, then $f_a(x)$ is zero at $x = 1$; and if $a = 0$, then $f_a(x)$ is zero at all the points in $[-1, 1]$.

A decimal value correct to the last decimal place defines an interval, just as 1.234~ defines the interval $1.234 \pm \frac{1}{2}$. A program that could determine x_0 to k or $k+1$ correct decimals would allow us to determine from a positive x_0 interval that $a \leq 0$, and from a negative x_0 interval that $a \geq 0$. From an x_0 interval that included the zero point, we could infer that a was zero. However, this contradicts Nonsolvable Problem 3.12, so the presumed program cannot exist.

We obtain a solvable problem if $f(x_0)$ is required only to be "close" to c:

SOLVABLE PROBLEM 6.2: Given a function $f(x)$ defined on an interval $[a, b]$ such that $f(a) \neq f(b)$, and given a number c between $f(a)$ and $f(b)$, find to k or $k+1$ correct decimals an argument x_0 in (a, b) such that $f(x_0) - c = 0.00\langle k \text{ zeros}\rangle 00\sim$.

Again for clarity, assume $f(a) < c < f(b)$. The other case $f(a) > c > f(b)$ would be treated analogously. We follow the same general procedure as described in the theorem, and construct two monotone sequences a_n and b_n with $f(a_n) < c < f(b_n)$, and such that $b_n - a_n = (b-a)/2^{n-1}$. Again $a_1 = a$ and $b_1 = b$, and assume the sequences are determined up to a_n and b_n, and let the point m_n be $(a_n + b_n)/2$, the midpoint of the interval $[a_n, b_n]$. This time we always obtain $D = k+1$ intervals for $f(m_n)$ and c, and, as long as these intervals do not intersect, follow the plan described in the theorem. If for some n, say N, the $D = k+1$ intervals for $f(m_N)$ and c intersect, then we have $|f(m_N) - c| < 4 \cdot 10^{-(k+1)}$ so $f(m_N) - c = 0.00\langle \text{k zeros}\rangle 00\sim$. We can compute m_N to either k or $k+1$ correct decimals, supply the result as x_0, and satisfy the conditions of this problem.

If we never find such intersecting intervals, then we continue the process until for some n, say N, the length of $[a_N, b_N]$ becomes less than $10^{-(k+2)}$. According to the theorem, a number x_0 such that $f(x_0) = c$ cannot fail to exist in $[a_N, b_N]$. Now we compute a $D = k+2$ interval for a_N (or for b_N), with the error bound of 1 unit in the last decimal place increased to 2 units to reflect the size of $[a_N, b_N]$. This approximation can be rounded to k places or $k+1$ places to give the appropriate x_0 decimal approximation. (The second paragraph after Solvable Problem 3.8 is relevant here.)

As Nonsolvable Problem 6.1 has shown, it is the possibility of $f(x)$ being constant over a subinterval of $[a, b]$ that makes the problem difficult. If $f(x)$

is known not to have any such subintervals, then we can obtain a better result:

SOLVABLE PROBLEM 6.3: Given a function $f(x)$ defined on an interval $[a, b]$ such that $f(a) \neq f(b)$, and such that there are no subintervals of $[a, b]$ on which $f(x)$ is constant, and given a number c between $f(a)$ and $f(b)$, find to k or $k+1$ correct decimals an argument x_0 in (a, b) such that $f(x_0) = c$.

Once more, assume $f(a) < c < f(b)$. The other case $f(a) > c > f(b)$ would be treated analogously. Here we adjust the definition of the two sequences a_n and b_n used in the proof of Theorem 6.8, so that their common limit is the desired number x_0, but we do not use the same procedure to obtain a_{n+1} and b_{n+1} from a_n and b_n. As before we take a_1 equal to a and take b_1 equal to b. To define a_{n+1} and b_{n+1} in terms of a_n and b_n, change m_n to be a number determined as follows: First obtain an approximating interval for $(a_n + b_n)/2$ that is sufficiently narrow that its overall length is less than $(b_n - a_n)/4$. Then make up a sequence $\widehat{r}_n$ of all the rational numbers that lie in this interval. (A similar procedure occurred in the proof of the Ceitin theorem.) Next compute c and the values $f(\widehat{r}_n)$ to increasing accuracy, stopping the computation as soon as we find an interval for some $f(\widehat{r}_{n_0})$ that does not intersect an interval for c. Because $f(x)$ cannot be constant on any subinterval of $[a, b]$, eventually a rational $\widehat{r}_{n_0}$ must be found. We choose m_n equal to the rational number $\widehat{r}_{n_0}$, and choose a_{n+1} and b_{n+1} in the same way as before, and then have the inequality $b_{n+1} - a_{n+1} < (\frac{3}{4})(b_n - a_n)$. The sequences a_n and b_n are monotone and $b_n - a_n < (\frac{3}{4})^{n-1}(b - a)$, so eventually, when n becomes large enough, say N, we obtain $b_N - a_N < 10^{-(k+2)}$, and then a $D = k + 2$ interval approximation to a_N (or to b_N), with the error bound of 1 unit in the last decimal place increased to two units, can be rounded to k places or to $k + 1$ places to obtain the appropriate x_0 decimal approximation. (Again the second paragraph after Solvable Problem 3.8 is relevant here.)

6.8 Unbounded Continuous Functions on $[a, b]$

Even though the functions of computable calculus are continous at every point in their domain, these functions need not be bounded even when their domain is a finite closed interval.

THEOREM 6.9: There exists a function $t(x)$ defined over $[0, 1]$, which is unbounded over this interval.

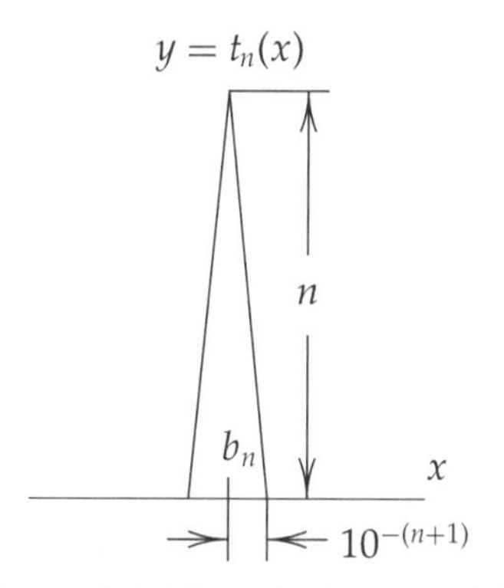

Figure 6.4. The spike function $t_n(x)$.

To define $t(x)$, let b_n be the sequence used to prove Specker's theorem, and for each term b_n define a "spike" function $t_n(x)$ by the equation

$$t_n(x) = n \cdot \max(0, 1 - |x - b_n| \cdot 10^{n+1})$$

The spike function $t_n(x)$ (see Fig. 6.4) has the value n at the point b_n, being nonzero in the interval $(b_n - 10^{-(n+1)}, b_n + 10^{-(n+1)})$ and zero everywhere else. The term b_n is separated from the succeeding term b_{n+1} by at least the distance $2 \cdot 10^{-(n+1)}$, so the nonzero domain of the spike function $t_n(x)$ does not intersect the nonzero domain of the spike function $t_{n+1}(x)$.

The function $t(x)$ equals $\sum_{n=1}^{\infty} t_n(x)$ and is defined on the interval $[0, 1]$. Let x be any number in $[0, 1]$ with decimal approximation function $x(D)$. Choosing a large D and obtaining $x(D) = 0.d_1d_2 \cdots d_D{}^{\pm}1$, if for some integer N less than D we find the digit d_N not equal to 2 or 7, then for $n > N$ we have $|x - b_n| > 10^{-(N+1)}$, so $t(x)$ may be computed as $\sum_{n=1}^{N} t_n(x)$. For the case where all the examined digits of the D decimal expansion equal 2 or 7, a different method is needed to determine the summation bound N. Here we may use the procedure described in Section 3.4 for constructing a number z to refute any supposed program $P^\star$ for deciding whether a real number is zero. We suppose that the given number x is the limit of the sequence b_n, and construct a program $P^\star$ that determines whether any number a equals zero by examining the appropriate digit of x's decimal expansion. This program will give the wrong answer for a constructible number z with approximation program $z(D)$ supplied by a program P_z, and now we proceed as follows: For N equal to $<P_z>$, the digit d_N of x is examined to determine whether it is 2 or 7. If it is neither, then we proceed as before. If the digit d_N equals 7, then we may use N as our summation bound because it is certain that $z = 0$, and the corresponding digit $d_N^{(n)}$ of any b_n is always 2. And if d_N equals 2, then it is certain that $z \neq 0$, and an integer N' may be computed such that $d_N^{(n)}$ first becomes 7 for $n = N'$. Now we use N' as our summation bound.

6.9 Limits of Sequences of Functions

The concepts of convergent sequences and Cauchy sequences are easily generalized to apply to sequences of functions.

DEFINITION 6.7: *A sequence of functions $f_n(x)$ converges to a limit $f(x)$ on an interval I,*

$$\text{written } \lim_{n\to\infty} f_n(x) = f(x) \quad \text{for } x \text{ in } I$$

if both $f_n(x)$ and $f(x)$ are defined on I, and there is a semifunction $N(\epsilon, x)$, defined for positive ϵ and x in I, such that for any number x_0 in I,

$$n > N(\epsilon, x_0) \quad \textit{implies} \quad |f_n(x_0) - f(x_0)| < \epsilon$$

If $N(\epsilon, x)$ can be replaced by $N(\epsilon)$, then $f_n(x)$ is said to converge uniformly *to $f(x)$ on I.*

When $f_n(x)$ has the form

$$\sum_{i=1}^{n} h_i(x)$$

it is customary to denote the limit function by the notation

$$\sum_{i=1}^{\infty} h_i(x)$$

A Cauchy sequence of functions has this definition:

DEFINITION 6.8: *A sequence of functions $f_n(x)$ is a Cauchy sequence of functions on an interval I, if $f_n(x)$ is defined on I, and there is a semifunction $N_C(\epsilon, x)$, defined for ϵ positive and x in I, such that for any number x_0 in I,*

$$n, m > N_C(\epsilon, x_0) \quad \textit{implies} \quad |f_n(x_0) - f_m(x_0)| < \epsilon$$

If $N_C(\epsilon, x)$ can be replaced by a semifunction $N_C(\epsilon)$, then $f_n(x)$ is said to be a uniform Cauchy sequence of functions *on I.*

In computable calculus, because we cannot use the result that a monotone bounded sequence converges to a limit, the concept of a Cauchy sequence of functions comes into play as an alternate way of obtaining standard results. The next theorem is analogous to Theorem 6.1, which concerns sequences, and has many applications.

THEOREM 6.10: A sequence of functions $f_n(x)$ converges on an interval I if and only if it is a Cauchy sequence of functions on I. Moreover, $f_n(x)$ converges uniformly on I if and only if it is a uniform Cauchy sequence of functions on I.

Suppose $f_n(x)$ converges to a function $f(x)$. Then we have a semifunction $N(\epsilon, x)$ such that for any x_0 in I,

$$n > N(\epsilon, x_0) \quad \text{implies} \quad |f_n(x_0) - f(x_0)| < \epsilon$$

If $n, m > N(\epsilon/2, x_0)$, we have

$$\begin{aligned}|f_n(x_0) - f_m(x_0)| &= |f_n(x_0) - f(x_0) + f(x_0) - f_m(x_0)| \\ &\leq |f_n(x_0) - f(x_0)| + |f(x_0) - f_m(x_0)| < \frac{\epsilon}{2} + \frac{\epsilon}{2} = \epsilon\end{aligned}$$

We can take $N_C(\epsilon, x)$ equal to $N(\epsilon/2, x)$, and $f_n(x)$ is a Cauchy sequence of functions.

Conversely, if $f_n(x)$ is a Cauchy sequence of functions, we must show how to obtain a limit function $f(x)$ as well as the semifunction $N(\epsilon, x)$. As in the proof of Theorem 6.1, it is convenient here to assume that the approximation algorithm for x is of type E, with the returned $f(x)$ algorithm also being of type E. The returned $y(E)$ algorithm chooses a positive integer $n_{x,E}$ larger than $N_C(E/2, x)$, and takes $y(E)$ equal to an $E/2$ approximation interval to $f_{n_{x,E}}(x)$, except that the error bound is increased by $E/2$. As before, this gives $y(E)$ the property that for any E_1 and E_2 the intervals $y(E_1)$ and $y(E_2)$ intersect, so $y(E)$ defines a real number. Also, as before, we take $N(\epsilon, x)$ equal to $N_C(\epsilon/2, x)$ and find that for $n > N(\epsilon, x)$ we have $|f_n(x) - y| < \epsilon$.

This system of assigning a limit value for each number x given by an approximation algorithm $x(E)$ ensures that at least a semifunction $f(x)$ has been defined. To show that $f(x)$ is actually a function, let $x'(E)$ be another approximation algorithm with $x = x'$. The corresponding y' approximation algorithm is similarly obtained by choosing a positive integer $n_{x',E}$ larger than $N_C(E/2, x')$, and taking $y'(E)$ equal to an $E/2$ approximation interval to $f_{n_{x',E}}(x')$, except that the error bound is increased by $E/2$. We must show that for any E_1 and E_2, the intervals $y(E_1)$ and $y'(E_2)$ intersect. Here, if $n_{x,E_1} \geq n_{x',E_2}$, we have $|f_{n_{x,E_1}}(x) - f_{n_{x',E_2}}(x')| < E_2/2$, and if $n_{x,E_1} < n_{x',E_2}$, we have $|f_{n_{x,E_1}}(x) - f_{n_{x',E_2}}(x')| < E_1/2$. In any case the two intervals $y(E_1)$ and $y'(E_2)$ are sure to intersect, so a function is defined.

The special cases of uniformly convergent sequences of functions and uniform Cauchy sequences of functions are treated in analogous fashion.

COROLLARY: Let a_n be a sequence of positive numbers such that the sequence

$$b_n = \sum_{k=1}^{n} a_k$$

converges. If $h_n(x)$ is a sequence of functions defined on an interval I, such that $|h_n(x)| \leq a_n$, then the sequence of functions

$$f_n(x) = \sum_{k=1}^{n} h_k(x)$$

converges uniformly on I to a function $f(x)$.

The sequence b_n is a Cauchy sequence, so there is a semifunction $N_C(\epsilon)$ such that $n, m > N_C(\epsilon)$ implies $|b_n - b_m| < \epsilon$. Then the sequence of functions $f_n(x)$ is a uniform Cauchy sequence of functions, because if $n, m > N_C(\epsilon)$, then for any x_0 in I we have

$$\begin{aligned} |f_n(x_0) - f_m(x_0)| &= \left| \sum_{k=\min(n,m)+1}^{\max(n,m)} h_k(x_0) \right| \\ &\leq \sum_{k=\min(n,m)+1}^{\max(n,m)} |h_k(x_0)| \\ &\leq \sum_{k=\min(n,m)+1}^{\max(n,m)} a_k = |b_n - b_m| < \epsilon \end{aligned}$$

The theorem now implies that $f_n(x)$ converges uniformly on I to a limit function $f(x)$.

6.10 Limits of Functions with More Than One Variable

Now that functions $f(x_1, \ldots, x_n)$ of two or more intervals are defined, we can also define limits for such functions. Some new notation is needed here.

DEFINITION 6.9: *The difference between two points* $(x_1, \ldots, x_n)$ *and* $(y_1, \ldots, y_n)$ *is*

$$(x_1, \ldots, x_n) - (y_1, \ldots, y_n) = (x_1 - y_1, \ldots, x_n - y_n).$$

The quantity $|(x_1, \ldots, x_n)|$ *is defined by the equation*

$$|(x_1, \ldots, x_n)| = \sqrt{x_1^2 + \ldots + x_n^2}$$

We measure the separation between two argument points $(x_1, \ldots, x_n)$ and $(x'_1, \ldots, x'_n)$ of a function $f(x_1, \ldots, x_n)$ by the quantity

$$|(x_1, \ldots, x_n) - (x'_1, \ldots, x'_n)| = \sqrt{(x_1 - x'_1)^2 + \ldots + (x_n - x'_n)^2}$$

Notice for the case $n = 1$, this measure gives the standard measure $|x_1 - x'_1|$. A *neighborhood* of $(x_1, \ldots, x_n)$ consists of the points $(x'_1, \ldots, x'_n)$ such that $|(x_1, \ldots, x_n) - (x'_1, \ldots, x'_n)| < \epsilon$ for some positive number ϵ. Now we are able to define a limit for functions of more than one variable:

DEFINITION 6.10: *A function* $f(x_1, \ldots, x_n)$ *has a limit* L *as* $(x_1, \ldots, x_n)$ *approaches a point* $(c_1, \ldots, c_n)$*,*

$$\text{written} \lim_{(x_1,\ldots,x_n)\to(c_1,\ldots,c_n)} f(x_1, \ldots, x_n) = L$$

if the domain of $f(x_1, \ldots, x_n)$ *includes a bounded region* $I^{(n)}$ *that either contains* $(c_1, \ldots, c_n)$ *or has this point as a boundary point, and there is a positive-valued semifunction* $\delta(\epsilon)$ *of the positive argument* ϵ*, such that for any point* $(x_1, \ldots, x_n)$ *for which* f *is defined,*

$$|(x_1, \ldots, x_n) - (c_1, \ldots, c_n)| < \delta(\epsilon) \quad \textit{implies} \quad |f(x_1, \ldots, x_n) - L| < \epsilon$$

For this limit to exist, $f(x_1, \ldots, x_n)$ must be defined at least in some bounded region $I^{(n)}$ having $(c_1, \ldots, c_n)$ as a boundary point, such a domain providing sufficient neighboring points to $(c_1, \ldots, c_n)$ for the limit concept to make sense. However, the requirement

$$|(x_1, \ldots, x_n) - (c_1, \ldots, c_n)| < \delta(\epsilon) \quad \text{implies} \quad |f(x_1, \ldots, x_n)) - L| < \epsilon$$

must hold for *all* points $(x_1, \ldots, x_n)$ where f is defined and not just for the points in the minimum domain $I^{(n)}$.

Limits where the modulus function δ depends on other variables besides the variable ϵ occur frequently when considering derivatives and partial derivatives. The only change needed in our limit definition to cover such cases is to replace the limit modulus $\delta(\epsilon)$ by a modulus with additional variables. For instance, the conventional calculus definition of derivative is

$$\lim_{h\to 0} \frac{f(x+h) - f(x)}{h} = f'(x)$$

If the function $f(x)$ and its derivative are defined in some interval I, then this limit requires a limit modulus $\delta(\epsilon, x)$ defined for ϵ positive and x in I.

A major use of limits in computable calculus is in extending the domain of a function to include points where the function previously was undefined. Two results in this direction are given next.

THEOREM 6.11: If $f(x_1, \ldots, x_n)$ is defined on a region $I^{(n)}$ except for one point $(c_1, \ldots, c_n)$, and $\lim_{(x_1,\ldots,x_n)\to(c_1,\ldots,c_n)} f(x_1, \ldots, x_n) = L$, then $f(x_1, \ldots, x_n)$ can be extended to become a function defined everywhere on $I^{(n)}$.

The extended function defined on $I^{(n)}$ is

$$f_1(x_1, \ldots, x_n) = \begin{cases} f(x_1, \ldots, x_n) & \text{if } (x_1, \ldots, x_n) \neq (c_1, \ldots, c_n) \\ L & \text{if } (x_1, \ldots, x_n) = (c_1, \ldots, c_n) \end{cases}$$

The f_1 program, with input $x_1(K), \ldots, x_n(K)$ defining a point $(x_1, \ldots, x_n)$ in $I^{(n)}$, composes and returns an approximation algorithm $y(K)$ that uses as subroutines the f program, the programs $x_i(K)$, the programs $c_i(K)$, the program $L(K)$ defining L, and the $\delta(\epsilon)$ program. The approximation algorithm $y(K)$ for any input K first computes $L(K) = m \pm e$ with the tentative goal of returning $m \pm e'$ with e' larger than e. Let d be the positive number $d = \delta(e)$. The program $y(K)$ forms more and more accurate approximations to d and to $|(x_1, \ldots, x_n) - (c_1, \ldots, c_n)|$, and eventually, according to Solvable Problem 3.10, it can decide for the nonnegative number $|(x_1, \ldots, x_n) - (c_1, \ldots, c_n)|$ whether it is >0, $>d$, or $<d$. In the first and second cases, the program abandons its tentative goal, obtains $\widehat{y}(K)$ by calling the f subroutine supplied with x_i arguments, then computes and returns $\widehat{y}(K)$. In the remaining case, it satisfies the tentative goal by returning $m \pm 2e$ to reflect the inequality

$$|(x_1, \ldots, x_n) - (c_1, \ldots, c_n)| < d = \delta(e)$$

In this way, for every K a value for $y(K)$ is obtained that satisfies the intersection requirement $y(K_1) \doteq y(K_2)$.

The next theorem is used in Chapter 8.

THEOREM 6.12: If $f(x, y)$ is defined for x in some interval I_1 and y in some interval I_2, except for the points in this xy region that lie on a line $y = mx + b$, and there is a function $g(x)$ such that for every x_0 in I_1 we have $\lim_{(x,y)\to(x_0, mx_0+b)} f(x, y) = g(x_0)$, then $f(x, y)$ can be extended to become a function defined everywhere in the xy region.

In the xy region the extended function is

$$f_1(x, y) = \begin{cases} f(x, y) & \text{if } y \neq mx + b \\ g(x) & \text{if } y = mx + b \end{cases}$$

For the limit specified in the theorem there is a modulus $\delta(\epsilon, x)$ defined for ϵ positive and x in I_1, such that $|(x, y) - (x_0, mx_0 + b)| < \delta(\epsilon, x_0)$ implies $|f(x, y) - g(x_0)| < \epsilon$. The f_1 program, given the input $x(K)$ defining a number x in I_1 and the input $y(K)$ defining a number y in I_2, composes and returns an approximation algorithm $u(K)$ that uses as subroutines the f program, the g program, the δ program, and the various approximation algorithms $x(K)$, $y(K)$, $m(K)$ and $b(K)$. For any input K, the algorithm $u(K)$ first forms $g(x)(K) = m \pm e$ by calling the g subroutine and evaluating the returned approximation algorithm, with the tentative goal of returning the interval $m \pm e'$ with $e' > e$. Let d be the positive number $\delta(e, x)$. The program $u(K)$ forms more and more accurate approximations to d and to $|y - mx - b| = |(x, y) - (x, mx + b)|$ until it determines a true inequality from the three possibilities: $|y - mx - b| > 0, |y - mx - b| < d, |y - mx - b| > d$. (Here again we are making use of Solvable Problem 3.10.) In the first and last case, the program $u(K)$ abandons its tentative goal, obtains $\widehat{u}(K)$ by calling the f subroutine with the $x(K)$ and $y(K)$ input, and then returns $\widehat{u}(K)$. In the middle case, it satisfies its tentative goal and returns $m \pm 2e$ to reflect the inequality $|y - mx - b| < d = \delta(e, x)$. In this way, for every K a value for $u(K)$ is obtained that satisfies the intersection requirement $u(K_1) \doteq u(K_2)$.

CHAPTER 7

Uniformly Continuous Functions

7.1 Introduction

In the preceding chapter we encountered certain functions $f(x)$ that have no counterpart in conventional calculus. An example is the function $t(x)$ of Theorem 6.9, which is defined and continuous at each point of $[0, 1]$, but which is unbounded on this interval. We can obtain functions more like the functions usually employed in an elementary calculus course by requiring the functions to be uniformly continuous. We copy from conventional calculus the uniform continuity definition:

DEFINITION 7.1: *A function $f(x)$ defined on an interval I is uniformly continuous there if there is a positive-valued semifunction $\delta(\epsilon)$ of the positive argument ϵ, such that for any numbers x_1, x_2 in I,*

$$|x_1 - x_2| < \delta(\epsilon) \quad \textit{implies} \quad |f(x_1) - f(x_2)| < \epsilon$$

If $f(x)$ is defined on a finite interval I that is of the form $[a, b)$ or (a, b) or $(a, b]$, then, as is well known, $f(x)$ may not be uniformly continuous on I. An example of this would be the function $1/x$ defined on any finite interval with a zero endpoint, the domain of course not including the point zero. In conventional calculus, if $f(x)$ is defined on a closed interval $[a, b]$ and is continuous at all points of the interval, then $f(x)$ is uniformly continuous there. We do not obtain this result in computable calculus, as Theorem 6.9 has shown. The function $t(x)$ of that theorem is defined and continuous on $[0, 1]$, but is not uniformly continuous on $[0, 1]$ because of the arbitrarily sharp spikes it has there.

In computable calculus, a function uniformly continuous on a closed interval I has many (but not all) of the properties expected in conventional calculus for a function continuous over that domain, and so these functions may be considered the replacements for the conventional calculus functions that are continuous at each point of I.

The next theorem shows that uniform continuity is preserved when uniform continuous functions are combined in standard ways to form new functions.

THEOREM 7.1: If $f(x)$ and $g(x)$ are uniformly continuous functions over a finite closed interval $I = [a, b]$, then the functions $|f(x)|$, $f(x) + g(x)$, $f(x) - g(x)$, $f(x)g(x)$, $\max(f(x), g(x))$, $\min(f(x), g(x))$, and $\sqrt[n]{f(x)}$ for n odd, are also uniformly continuous over I. If $f(x) \geq 0$ on I, then the function $\sqrt[n]{f(x)}$ for n even is uniformly continuous on I. If for $g(x)$ on I we have the bound $|g(x)| > e_g$, where e_g is a positive number, then the function $f(x)/g(x)$ is uniformly continuous over I. If $h(x)$ is a uniformly continuous function defined on a closed interval I_1, and if for all x in I, the function $f(x)$ has a value in I_1, then the function $h(f(x))$ is uniformly continuous in I.

It is easy to see that the function x and any constant function both are uniformly continuous over any closed interval $[a, b]$, because the uniform continuity modulus $\delta(\epsilon) = \epsilon$ serves for them. After this theorem is proved, any polynomial function is uniformly continuous over any finite closed interval, because with a finite number of the operations listed, the polynomial can be constructed out of constant functions and the x function.

Theorem 4.1 showed that the various constructions mentioned in this theorem are functions, so to prove this theorem we need only show for each function type how to obtain a uniform continuity modulus from the moduli $\delta_f(\epsilon)$ for $f(x)$, $\delta_g(\epsilon)$ for $g(x)$, and $\delta_h(\epsilon)$ for $h(x)$. First we note that we can find an upper bound M_f for $|f(x)|$ on I by choosing some convenient ϵ, say ϵ_0, finding an integer N such that $(b - a)/N < \delta_f(\epsilon_0)$, and then taking M_f equal to $\epsilon_0 + \max_{i=0}^{N} |f(a + i(b - a)/N)|$. The number M_f is an upper bound, because if x is any point in $[a, b]$, then x is in some subinterval

$[a + i_0(b-a)/N, a + (i_0+1)(b-a)/N]$, so

$$
\begin{aligned}
|f(x)| &= \left| f(x) - f\left(a + i_0\frac{b-a}{N}\right) + f\left(a + i_0\frac{b-a}{N}\right)\right| \\
&\le \left| f(x) - f\left(a + i_0\frac{b-a}{N}\right)\right| + \left| f\left(a + i_0\frac{b-a}{N}\right)\right| \\
&< \epsilon_0 + \left| f\left(a + i_0\frac{b-a}{N}\right)\right| \\
&\le \epsilon_0 + \max_{i=0}^{N}\left| f\left(a + i\frac{b-a}{N}\right)\right| = M_f
\end{aligned}
$$

Similarly, an upper bound M_g can be found for $|g(x)|$ over I.

Here are the moduli for the various cases:

Case	**Modulus**
$\|f(x)\|$	$\delta_f(\epsilon)$
$f(x)+g(x)$	$\min\left(\delta_f\left(\frac{\epsilon}{2}\right), \delta_g\left(\frac{\epsilon}{2}\right)\right)$
$f(x)-g(x)$	$\min\left(\delta_f\left(\frac{\epsilon}{2}\right), \delta_g\left(\frac{\epsilon}{2}\right)\right)$
$f(x)g(x)$	$\min\left(\delta_f\left(\frac{\epsilon}{2M_g}\right), \delta_g\left(\frac{\epsilon}{2M_f}\right)\right)$
$\max(f(x), g(x))$	$\min(\delta_f(\epsilon), \delta_g(\epsilon))$
$\min(f(x), g(x))$	$\min(\delta_f(\epsilon), \delta_g(\epsilon))$
$\sqrt[n]{f(x)}$	$\delta_f\left(\left(\frac{\epsilon}{2}\right)^n\right)$
$f(x)/g(x)$	$\min\left(\delta_f\left(\frac{\epsilon e_g^2}{2M_g}\right), \delta_g\left(\frac{\epsilon e_g^2}{2M_f}\right)\right)$
$h(f(x))$	$\delta_f(\delta_h(\epsilon))$

Thus for the function $f(x)g(x)$, when $|x_1 - x_2|$ is less than the modulus shown, we have

$$
\begin{aligned}
|f(x_1)g(x_1) - f(x_2)g(x_2)| &= |f(x_1)g(x_1) - f(x_2)g(x_1) + f(x_2)g(x_1) - f(x_2)g(x_2)| \\
&\le |f(x_1)g(x_1) - f(x_2)g(x_1)| + |f(x_2)g(x_1) - f(x_2)g(x_2)| \\
&= |g(x_1)| \cdot |f(x_1) - f(x_2)| + |f(x_2)| \cdot |g(x_1) - g(x_2)| \\
&< M_g \cdot \frac{\epsilon}{2M_g} + M_f \cdot \frac{\epsilon}{2M_f} = \frac{\epsilon}{2} + \frac{\epsilon}{2} = \epsilon
\end{aligned}
$$

For the function $f(x)/g(x)$, when $|x_1 - x_2|$ is less than the modulus shown, we have

$$\begin{aligned}\left|\frac{f(x_1)}{g(x_1)} - \frac{f(x_2)}{g(x_2)}\right| &= \left|\frac{f(x_1)g(x_2) - g(x_1)f(x_2)}{g(x_1)g(x_2)}\right| \\ &= \frac{|f(x_1)g(x_2) - f(x_2)g(x_2) + f(x_2)g(x_2) - g(x_1)f(x_2)|}{|g(x_1)g(x_2)|} \\ &\le \frac{|g(x_2)| \cdot |f(x_1) - f(x_2)| + |f(x_2)| \cdot |g(x_2) - g(x_1)|}{|g(x_1)| \cdot |g(x_2)|} \\ &< \frac{M_g \cdot \frac{\epsilon\epsilon_g^2}{2M_g} + M_f \cdot \frac{\epsilon\epsilon_g^2}{2M_f}}{\epsilon_g \epsilon_g} = \frac{\epsilon}{2} + \frac{\epsilon}{2} = \epsilon\end{aligned}$$

For the function $\sqrt[n]{f(x)}$, when $|x_1 - x_2|$ is less than the modulus shown, we consider two cases. If $|f(x_1)| < (\frac{\epsilon}{2})^2$ and $|f(x_1)| < (\frac{\epsilon}{2})^2$, then

$$\begin{aligned}|\sqrt[n]{f(x_1)} - \sqrt[n]{f(x_2)}| &\le |\sqrt[n]{f(x_1)}| + |\sqrt[n]{f(x_2)}| \\ &< \left|\sqrt[n]{\left(\frac{\epsilon}{2}\right)^n}\right| + \left|\sqrt[n]{\left(\frac{\epsilon}{2}\right)^n}\right| = \frac{\epsilon}{2} + \frac{\epsilon}{2} = \epsilon\end{aligned}$$

On the other hand, when either $|f(x_1)| \ge (\frac{\epsilon}{2})^n$ or $|f(x_2)| \ge (\frac{\epsilon}{2})^n$, then we use the identity

$$(\sqrt[n]{f(x_1)} - \sqrt[n]{f(x_2)}) \cdot \sum_{i=1}^{n} \left[(\sqrt[n]{f(x_1)})^{n-i}(\sqrt[n]{f(x_2)})^{i-1}\right] = f(x_1) - f(x_2)$$

If $f(x_1)$ and $f(x_2)$ are both nonnegative or both nonpositive (odd n), then we have

$$\begin{aligned}|\sqrt[n]{f(x_1)} - \sqrt[n]{f(x_2)}| &= \frac{|f(x_1) - f(x_2)|}{\left|\sum_{i=1}^{n}(\sqrt[n]{f(x_1)})^{n-i}(\sqrt[n]{f(x_2)})^{i-1}\right|} \\ &\le \frac{|f(x_1) - f(x_2)|}{(\sqrt[n]{(\frac{\epsilon}{2})^n})^{n-1}} < \frac{(\frac{\epsilon}{2})^n}{(\frac{\epsilon}{2})^{n-1}} = \frac{\epsilon}{2}\end{aligned}$$

If n is odd and $f(x_1)$ and $f(x_2)$ have opposite signs, Theorem 6.8 implies that between x_1 and x_2 a number x_3 cannot fail to exist such that $f(x_3) = 0$. We have

$$\begin{aligned}|\sqrt[n]{f(x_1)} - \sqrt[n]{f(x_2)}| &= |\sqrt[n]{f(x_1)} - \sqrt[n]{f(x_3)} + \sqrt[n]{f(x_3)} - \sqrt[n]{f(x_2)}| \\ &\le |\sqrt[n]{f(x_1)} - \sqrt[n]{f(x_3)}| + |\sqrt[n]{f(x_3)} - \sqrt[n]{f(x_2)}| \\ &< \frac{\epsilon}{2} + \frac{\epsilon}{2} = \epsilon\end{aligned}$$

Finally, for the last function $h(f(x))$, when $|x_1 - x_2|$ is less than the modulus shown, we have $|f(x_1) - f(x_2)| < \delta_h(\epsilon)$, which implies $|h(f(x_1)) - h(f(x_2))| < \epsilon$.

A sequence of functions also may be uniformly continuous:

DEFINITION 7.2: *A sequence of functions $f_n(x)$ is uniformly continuous on an interval I if for all n the term function $f_n(x)$ is defined on I, and there is a sequence of positive-valued semifunctions $\delta_n(\epsilon)$, such that for any numbers x_1, x_2 in I,*

$$|x_1 - x_2| < \delta_n(\epsilon) \quad \textit{implies} \quad |f_n(x_1) - f_n(x_2)| < \epsilon$$

If a sequence of functions $f_n(x)$ is uniformly continuous on an interval I and converges to a limit function $f(x)$, then $f(x)$ may or may not be uniformly continuous. An example where the limit function is not uniformly continuous is given by the function $t(x)$ which is the limit of the sequence of functions $\sum_{i=1}^{n} t_i(x)$ defined in the proof of Theorem 5.9. The next theorem gives a condition implying uniform continuity of the limit function.

THEOREM 7.2: If the sequence of functions $f_n(x)$ is uniformly continuous on an interval I and converges uniformly there to the function $f(x)$, then $f(x)$ is uniformly continuous on I.

Let $\delta_n(\epsilon)$ be the uniform continuity semifunction sequence for $f_n(x)$, and let $N(\epsilon)$ be the semifunction required for the uniform convergence of $f_n(x)$ to $f(x)$. The semifunction $\delta(\epsilon)$ needed for uniform continuity of $f(x)$ can be obtained this way: For any ϵ, we take $\delta(\epsilon)$ equal to $\delta_{N_0}(\epsilon/3)$ where N_0 is a positive integer larger than $N(\epsilon/3)$. Then for x_1, x_2 in I with $|x_1 - x_2| < \delta(\epsilon)$, we have

$$\begin{aligned} |f(x_1) - f(x_2)| &= \left|f(x_1) - f_{N_0}(x_1) + f_{N_0}(x_1) - f_{N_0}(x_2) + f_{N_0}(x_2) - f(x_2)\right| \\ &\leq \left|f(x_1) - f_{N_0}(x_1)\right| + \left|f_{N_0}(x_1) - f_{N_0}(x_2)\right| + \left|f_{N_0}(x_2) - f(x_2)\right| \\ &< \frac{\epsilon}{3} + \frac{\epsilon}{3} + \frac{\epsilon}{3} = \epsilon \end{aligned}$$

Consequently $f(x)$ is uniformly continuous on I.

COROLLARY: Let a_n be a sequence of positive numbers such that the sequence

$$b_n = \sum_{k=1}^{n} a_k$$

converges. If $h_n(x)$ is a uniformly continuous sequence of functions on an interval I, such that $|h_n(x)| \leq a_n$ for any x on I, then the sequence of functions

$$f_n(x) = \sum_{k=1}^{n} h_k(x)$$

converges uniformly on I to a uniformly continuous function $f(x)$.

Let $\delta_n(\epsilon)$ be the semifunction sequence needed for the uniform continuity of $h_n(x)$ on I. Then the semifunction sequence

$$\widehat{\delta}_n(\epsilon) = \min_{1 \leq k \leq n} \delta_k\left(\frac{\epsilon}{n}\right)$$

can serve as the uniform continuity moduli for $f_n(x)$ on I. By the Corollary to Theorem 6.10, the sequence of functions $f_n(x)$ converges uniformly on I to a limit function $f(x)$, which is uniformly continuous by the theorem just proved.

The Corollary can be used to show that various standard functions, such as e^x or $\sin(x)$, usually defined by infinite series, are uniformly continuous on any bounded closed interval.

7.2 Bounds of Uniformly Continuous Functions

The terminology about bounds previously used for sequences can also be applied to functions:

DEFINITION 7.3: *Let $f(x)$ be a function defined on an interval I. A number M is an* upper bound *for $f(x)$ over I if $f(x) \leq M$ for all x in I, and M is a* least upper bound *if the inequality $M \leq M'$ holds for any upper bound M'. For the case where there is an argument x_0 in I such that $f(x_0)$ equals its least upper bound M, then M is called the* maximum *of $f(x)$ on I. The terms lower bound, greatest lower bound, and minimum are defined analogously. The function $f(x)$ is* bounded *on I if it has an upper bound and a lower bound. In this case a number $\widehat{M}$ can be found such that that $|f(x)| \leq \widehat{M}$.*

The $t(x)$ example function of Theorem 6.9 shows that a function defined on a finite interval $[a, b]$, even though continuous at each point of the interval, may not be bounded there. This is not true for functions uniformly continuous over $[a, b]$.

THEOREM 7.3: If $f(x)$ is uniformly continuous on an interval $[a, b]$, then on this interval $f(x)$ is bounded, has a greatest lower bound m, and has a least upper bound M. If c is any number in the interval (m, M), that is, if $m < c < M$, then an argument x_0 in $[a, b]$ such that $f(x_0) = c$ cannot fail to exist.

We can define M as the limit of a certain sequence M_n. For any n, we choose an integer N_n such that $(b - a)/N_n < \delta(1/n)$, and then set M_n equal to $\max_{i=0}^{N_n} f(a + i(b - a)/N_n)$.

For any two integers n_1 and n_2, because each of the $f(x)$ arguments used to obtain M_{n_1} is $< \delta(1/n_2)$ away from an argument used to obtain M_{n_2}, we obtain the inequality

$$M_{n_1} = f\left(a + i_0\frac{b-a}{N_{n_1}}\right) < f\left(a + j_0\frac{b-a}{N_{n_2}}\right) + 1/n_2 \leq M_{n_2} + 1/n_2$$

so $M_{n_1} < M_{n_2} + 1/n_2$. Reversing the argument, we also obtain the inequality $M_{n_2} < M_{n_1} + 1/n_1$, and the two inequalities together imply

$$|M_{n_1} - M_{n_2}| < \max(1/n_1, 1/n_2) = 1/\min(n_1, n_2)$$

This implies that the sequence M_n is Cauchy and converges to a limit M. (The semifunction $N_C(\epsilon)$ can be $1/\epsilon$.) The number M is an upper bound for $f(x)$ over I, because if x_0 is any point of I, we have $f(x_0) < M_n + 1/n$, and taking the limit of this inequality, we obtain $f(x_0) \leq M$. Also, if M' is any upper bound for $f(x)$ on I, then $M_n \leq M'$ for all n, and taking the limit once more, we obtain $M \leq M'$. Thus M is a least upper bound. In similar fashion, a Cauchy sequence m_n may be defined, converging to a limit m, which is a greatest lower bound.

Now let c be a number satisfying $m < c < M$. By forming more and more accurate approximations to c and to more and more terms of the two sequences m_n and M_n, we could eventually find sequence terms $m_{n'}$ and $M_{n''}$ such that $m_{n'} < c < M_{n''}$. Next, by forming more and more accurate approximations to c and to all the function values used to form $m_{n'}$ and $M_{n''}$, we could eventually locate two x arguments x_1 and x_2 in I such that $f(x_1) < c < f(x_2)$. Now we can apply Theorem 5.8 and conclude that on the interval $[\min(x_1, x_2), \max(x_1, x_2)]$ an argument x_0 cannot fail to exist such that $f(x_0) = c$.

The functions studied in elementary calculus usually are uniformly continuous over every closed finite interval on which they are defined. In conventional calculus, a function uniformly continuous on $[a, b]$ assumes its least upper bound at some point in $[a, b]$, so a least upper bound is always a maximum function value. That this result does not go over into computable calculus was shown independently by both Zaslavsky [39] and Specker [35]:

THEOREM 7.4: There is a function $s(x)$ uniformly continuous on $[0, 1]$ that does not assume its least upper bound M; that is, $s(x) < M$ for all x in $[0, 1]$.

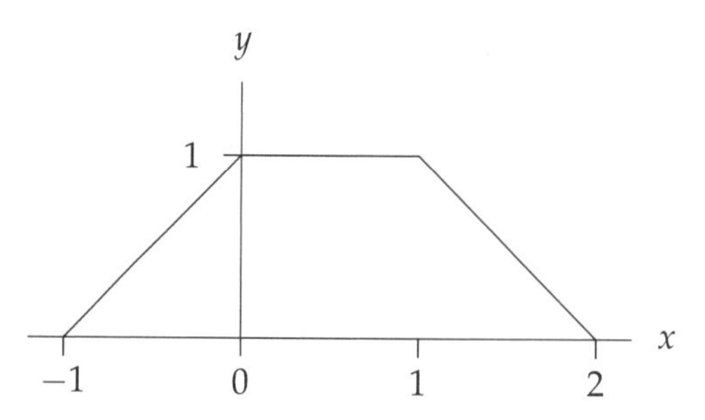

Figure 7.1. The function $w(x)$.

The function $w(x)$ used for building $s(x)$ is shown in Fig. 7.1. The function $w(x)$ is zero outside the interval $[-1, 2]$ and linear with slope 1 in the interval $[-1, 0]$, has the constant value 1 in the interval $[0, 1]$, and is linear with slope -1 in the interval $[1, 2]$. Note that the function $w(x - c)$ has the same graph as $w(x)$ except the upper constant region is shifted from the interval $[0, 1]$ to the interval $[c, c + 1]$.

To build $s(x)$ we use the sequence of functions $w_n(x)$, where $w_n(x)$ is a scaled version of $w(x)$ defined by equation $w_n(x) = 10^{-n}w(10^n x)$. The function $w_n(x)$ is zero outside the interval $[-10^{-n}, 2 \cdot 10^{-n}]$ and linear with slope 1 in the interval $[-10^{-n}, 0]$, has the constant value 10^{-n} in the interval $[0, 10^{-n}]$, and is linear with slope -1 in the interval $[10^{-n}, 2 \cdot 10^{-n}]$.

It is helpful here to recall the definition of the sequence b_n of Specker's theorem:

$$b_n = 0.d_1^{(n)}d_2^{(n)} \cdots d_n^{(n)}$$

Recall that the decimal digit $d_i^{(n)}$ takes the value 2 or 7. Similarly, for each positive integer n, we define the decimal form rationals $r_k^{(n)}$ which have the form

$$0.\widehat{d}_1^{(n)}\widehat{d}_2^{(n)} \cdots \widehat{d}_n^{(n)}$$

where each digit $\widehat{d}_i^{(n)}$ is allowed, with certain restrictions to be given, to equal either 2 or 7. Without the restrictions, this would mean that there are 2^n different rationals $r_k^{(n)}$, so k would run from 1 to 2^n. Recall that the digit $d_i^{(n)}$ equals 2 if, after following n steps of computation for $P_i(D)$ with D taken in succession to be $1, 2, \ldots$, it is not determined that the real number presumed defined by P_i is nonzero. The determination that this real number is nonzero is made by finding in n steps or less that some $P_i(D_0)$ output is an interval that is positive or negative. In this case, which we call the successful case, the digit $d_i^{(n)}$ equals 7. Here the restriction on the values allowed for $r_k^{(n)}$ is that digit $\widehat{d}_i^{(n)}$ may assume both values 2 and 7 after an n-step P_i computation is unsuccessful, but must take only one of these two values after a successful determination. If $P_i(D_0)$ is positive, the value taken must be 2, and if $P_i(D_0)$ is negative, the value taken must be 7. Thus, as more and more of the digits $d_i^{(n)}$ get their final value 7, more and more of the corresponding digits $\widehat{d}_i^{(n)}$ of $r_k^{(n)}$ get fixed in value, the fixed value being either 2 or 7.

We define the sequence of functions $s_n(x)$ by the equation

$$s_n(x) = \sum_k w_n(x - r_k^{(n)})$$

where the sum is over all allowable rationals $r_k^{(n)}$. Then the function $s(x)$ of the theorem is determined by the equation

$$s(x) = \sum_{n=1}^{\infty} s_n(x) = \lim_{N\to\infty} \sum_{n=1}^{N} s_n(x)$$

Each of the functions $s_n(x)$ is uniformly continuous with the same modulus function $\delta(\epsilon) = \epsilon$, because for any two of the summed functions $w_n(x - r_k^{(n)})$, the intervals where they are nonzero are disjoint. This also implies $|s_n(x)| \leq 10^{-n}$, so by the Corollary to Theorem 7.2, the function $s(x)$ is a uniformly continuous function. It is clear also that for all x in $[0, 1]$ the function $s(x)$ is bounded above by $\sum_{n=1}^{\infty} 10^{-n} = \frac{1}{9}$.

A particular function $w_n(x - r_{k_0}^{(n)})$ reaches its maximum value of 10^{-n} on some interval $I = [r_{k_0}^{(n)}, r_{k_0}^{(n)} + 10^{-n}]$, and the interval I contains two, one, or none of the rational points $r_k^{(n+1)}$, any such contained point being identical in decimal form to $r_{k_0}^{(n)}$ except for the last nonzero decimal digit of the point. The interval I contains one point if $\widehat{d}_{n+1}^{(n+1)}$ is fixed at 2 or fixed at 7, and I contains no point if for some positive integer $j_0 \leq n$ the digit $\widehat{d}_{j_0}^{(n+1)}$ becomes fixed to a value differing from the digit $\widehat{d}_{j_0}^{(n)}$ of $r_{k_0}^{(n)}$. Thus, the interval I contains two, one, or none of the subintervals on which a function $w_{n+1}(x - r_k^{n+1})$ is unequal to zero. At the point $x = r_{k_0}^{(n)}$, the function $s(x)$ is equal to $\sum_{i=1}^{n} 10^{-i}$, equivalent to the decimal value $0.11 \cdots 1$ with n digits of 1, so the least upper bound of $s(x)$ is definitely $\frac{1}{9}$.

Suppose we define a number x_0 in $[0, 1]$ by specifying a systematic way of choosing its successive decimal digits equal to 2 or 7, attempting to make the decimal expansion of x_0 to any number n of decimal digits the same as a number $r_k^{(n)}$. Perhaps we always choose 2, perhaps we alternate between 2 or 7, or perhaps we choose 2 or 7 dependent upon whether the nth decimal digit of some number such as π is or is not > 4. The function $s(x)$ cannot equal $\frac{1}{9}$ at any such point x_0 with approximation function $x_0(D)$, because with the help of the program $x_0(D)$ we would have a way of deciding for any real number a whether $a \geq 0$ or $0 \geq a$, contradicting Nonsolvable Problem 3.12. Given a real number a with approximation function $a(D)$ generated by a program P, we would be able to find the integer $i = \langle P \rangle$ associated with the program

P, and then we could evaluate $x_0(D)$ for $D = i + 1$ to determine whether the ith decimal digit of the approximant was 2, indicating that $a \geq 0$, or was 7, indicating that $0 \geq a$.

Thus functions uniformly continuous on $[a, b]$ may not have a maximum value on the interval. Of course if we know that on $[a, b]$ the function $f(x)$ is either monotone increasing or monotone decreasing, then it must have a maximum value.

DEFINITION 7.4: *A function $f(x)$ is* monotone increasing *on an interval I if $f(x)$ is defined on this interval, and if for any two numbers x_1, x_2 in I,*

$$x_1 < x_2 \quad \textit{implies} \quad f(x_1) \leq f(x_2)$$

The function $f(x)$ is strictly monotone increasing *if in the preceding display text the relation $\leq$ can be replaced by $<$. A* monotone decreasing *function and a* strictly monotone decreasing *function are defined analogously. A function* monotone *on I is either monotone increasing or monotone decreasing. A function* strictly monotone *on I is either strictly monotone increasing or strictly monotone decreasing.*

Let us suppose that the domain interval I of a function can be divided into a finite number of subintervals on each of which $f(x)$ is either monotone increasing or monotone decreasing. Let us call a function for which this division is possible *piecewise monotone* on I. Here is a more formal definition for the case where I is $[a, b]$:

DEFINITION 7.5: *A function $f(x)$ defined on an interval $[a, b]$ is piecewise monotone on $[a, b]$ if there is a finite sequence x_i, such that $x_0 = a$ and $x_N = b$, with $x_{i-1} < x_i$ for $1 \leq i \leq N$, and such that $f(x)$ is monotone on each subinterval $[x_{i-1}, x_i]$.*

For instance the function

$$g(x) = \begin{cases} x \sin x & \text{if } x \neq 0 \\ 0 & \text{if } x = 0 \end{cases}$$

is not piecewise monotone on [0, 1], but every algebraic function is piecewise monotone on any interval $[a, b]$ where it is defined. A function that is piecewise monotone on $[a, b]$ must be uniformly continuous on $[a, b]$, and also must have a maximum on this interval, certain to be assumed at one of the subdivision points x_i.

Often for a function $f(x)$, we may know it to be piecewise monotone on a certain interval $[a, b]$, but may not actually have at hand the finite sequence x_i that divides $[a, b]$ into appropriate subintervals. We describe such functions as "known to be piecewise monotone on $[a, b]$." We consider now whether for functions like these, which have a maximum on $[a, b]$, we can find an accurate approximation to a point in $[a, b]$ where this maximum is assumed.

NONSOLVABLE PROBLEM 7.1: Given a function $f(x)$ uniformly continuous and known to be piecewise monotone on an interval $[a, b]$, find to k or $k+1$ correct decimals a point x_0 in $[a, b]$ where $f(x_0)$ equals the maximum of $f(x)$ on $[a, b]$.

Take the interval $[a, b]$ to be $[-1, 1]$, and consider the function $f_c(x)$ with parameter c defined by the equation $f_c(x) = cx$. If $c > 0$, the maximum value for this function occurs at $x = 1$; if $c < 0$, the maximum value occurs at $x = -1$; and if $c = 0$, any point in $[-1, 1]$ can serve as a point defining the maximum value. If there were an ideal computer program which could solve this problem, then the supplied decimal value will either be positive, implying $c \geq 0$, negative, implying that $c \leq 0$, or zero, implying $c = 0$. From the value returned, we can solve Nonsolvable Problem 3.12 for any number c, a contradiction. Hence the problem is nonsolvable. For our function $f_c(x)$, we know the maximum occurs at the endpoint -1 or at the endpoint $+1$, but nevertheless sometimes have difficulty deciding which endpoint has the maximum value.

To convert the problem to a solvable one, we give up finding with certainty the point where the maximum is attained, and just find a point that matches to D decimals the maximum value.

SOLVABLE PROBLEM 7.2: Given a function $f(x)$ uniformly continuous and known to be piecewise monotone on an interval $[a, b]$, find to k or $k+1$ correct decimals the maximum value of $f(x)$ over $[a, b]$, and find to k or $k+1$ correct decimals a point x_0 in $[a, b]$ such that $f(x_0)$ has a value identical to the displayed maximum value.

The problem does not specify that all such points be obtained, but only that one such point be obtained. If we repeat the solution with an increased k, we may locate a different point x_0.

The procedure here could be the following: Using the uniform continuity modulus $\delta(\epsilon)$, we find an integer N such that $(b-a)/N < \delta(10^{-(k+2)})$, and then have $N+1$ arguments x_i in $[a, b]$ defined by the equation

$$x_i = a + i\frac{b-a}{N} \quad \text{for} \quad i = 0, 1, \ldots, N$$

For each x_i we obtain a $D = k+2$ interval for $f(x_i)$ with error bound 1 unit in the last decimal place. We choose an argument x_{i_0} whose f decimal approximant is the largest, and after increasing its f error bound by another unit in the last place, to represent the $10^{-(k+2)}$ argument that was used with $\delta(\epsilon)$, we now have an interval for the $f(x)$ maximum over $[a, b]$. As we noted in Section 3.6, a decimal number with $k + 2$ decimals having an error bound of 2 units in its last place allows k or $k + 1$ correct decimal places for the number to be obtained. The $f(x)$ maximum may now be displayed to k or $k + 1$ correct decimals, and then the argument x_{i_0} may be displayed to the same accuracy to satisfy the conditions of the problem.

CHAPTER

The Derivative

8.1 A Difficulty with the Derivative Definition

We take the conventional definition of derivative as our starting definition:

TENTATIVE DEFINITION 8.1: *The derivative of a function $f(x)$ at $x = x_0$ is a real number, to be denoted by the symbol $f'(x_0)$, provided the limit given here exists:*

$$\lim_{h \to 0} \frac{f(x_0 + h) - f(x_0)}{h} = f'(x_0)$$

Suppose the function $f(x)$ is defined on an interval I. The derivative of $f(x)$ is a function $f'(x)$ defined on I, provided the limit given here exists for any x_0 in I:

$$\lim_{h \to 0} \frac{f(x_0 + h) - f(x_0)}{h} = f'(x_0)$$

We have labeled this definition as tentative, because it sometimes causes the derivative $f'(x)$ of a continuous function $f(x)$ to be discontinuous at a point in the domain of $f(x)$. This flaw, first noted by Zaslavsky [44], is serious in computable calculus because a discontinuous derivative function cannot be realized at the point of discontinuity. A simple example of this trouble occurs with the function

$$s(x) = \begin{cases} x^2 \sin(\frac{1}{x}) & \text{if } x \neq 0 \\ 0 & \text{if } x = 0 \end{cases} \tag{8.1}$$

According to Theorem 6.7, $s(x)$ is a valid function defined everywhere, because $\lim_{x\to 0} x^2 \sin(1/x) = 0$.

The derivative of $x^2\sin(1/x)$ for $x \neq 0$ is $2x\sin(1/x) - \cos(1/x)$. This $s(x)$ derivative, defined for $x \neq 0$, cannot be extended to become a function defined for $x = 0$, because $\lim_{x\to 0}\ (2x\sin(1/x) - \cos(1/x))$ does not exist. On the other hand, $s'(0)$ is defined and equals 0 because

$$\lim_{h\to 0} \frac{s(0+h) - s(0)}{h} = \lim_{h\to 0} \frac{h^2\sin(1/h) - 0}{h} = \lim_{h\to 0} h\sin\left(\frac{1}{h}\right) = 0$$

We need to make the derivative definition a little more restrictive, so that this difficulty does not occur. Our definition then is

FINAL DEFINITION 8.1: *The derivative of a function $f(x)$ at $x = x_0$ is a real number, to be denoted by the symbol $f'(x_0)$, provided the limit given here exists:*

$$\lim_{(x,h)\to(x_0,0)} \frac{f(x+h) - f(x)}{h} = f'(x_0)$$

Suppose the function $f(x)$ is defined on an interval I. The derivative of $f(x)$ is a function $f'(x)$ defined on I, provided the limit given here exists for any point x_0 in I:

$$\lim_{(x,h)\to(x_0,0)} \frac{f(x+h) - f(x)}{h} = f'(x_0)$$

Figure 8.1 shows the δ neighborhoods for obtaining an ϵ bound for both the tentative definition and the final one. Note that the function $g(x,h) = (f(x+h) - f(x))/h$ is not defined in the xh plane on the line $h = 0$. For

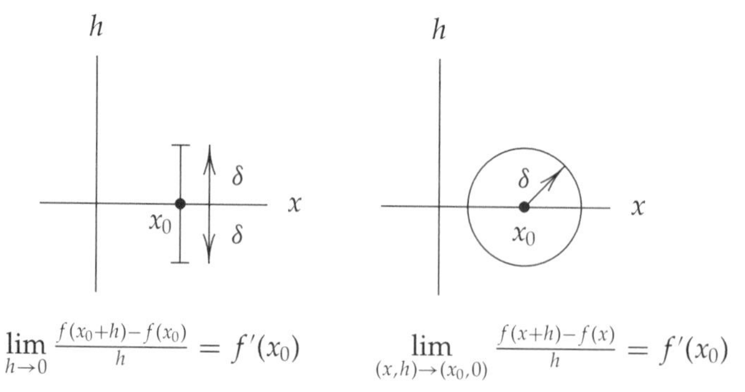

Figure 8.1. The δ neighborhood in (x, h) space needed to bound two derivative quotients within ϵ of $f'(x_0)$.

$\lim_{(x,h)\to(x_0,0)}(f(x+h)-f(x))/h$, the conventional limit definition requires $g(x,h)=(f(x+h)-f(x))/h$ to be defined in a punctured neighborhood of $(x_0,0)$, which is not possible in this case. Thus our relaxed limit definition is needed in order for this limit to make sense. In addition, for the case where I is $[a,b]$, the derivative $f'(a)$, if it exists, is obtained by a right-hand limit, and $f'(b)$, if it exists, is obtained by a left-hand limit. Thus, the final derivative definition includes the special case of derivatives at endpoints of domain intervals, whereas if we use the conventional limit definition, such derivatives require a separate definition.

Using the revised definition, we recompute the derivative of $s(x)$ at $x=0$ by first noting that for $(x+h)\neq 0$ and $x\neq 0$, we have

$$\begin{aligned}
\frac{s(x+h)-s(x)}{h} &= \frac{(x+h)^2\sin(1/(x+h))-x^2\sin(1/x)}{h} \\
&= \frac{x^2(\sin(1/(x+h))-\sin(1/x))}{h} \\
&\quad + \left(2x\sin\left(\frac{1}{x+h}\right)+h\sin\left(\frac{1}{x+h}\right)\right) \\
&= \frac{x^2}{h}\cdot\left(\sin\left(\frac{1}{x+h}\right)-\sin\left(\frac{1}{x}\right)\right) \\
&\quad + \left(2x\sin\left(\frac{1}{x+h}\right)+h\sin\left(\frac{1}{x+h}\right)\right)
\end{aligned}$$

The term $2x\sin(1/(x+h))+h\sin(1/(x+h))$ approaches 0 as (x,h) approaches $(0,0)$, but the preceding term can be made arbitrarily large or arbitrarily small in any neighborhood of $(0,0)$, so $\lim_{(x,h)\to(0,0)}(s(x+h)-s(x))/h$ is not defined.

As will be clear shortly, with this derivative definition we still obtain all the standard rules of differentiation. Thus, with this derivative definition, for the function $s(x)$ we still obtain for $x\neq 0$ the derivative function $s'(x)=2x\sin(1/x)-\cos(1/x)$, which clearly cannot be extended to include the argument $x=0$. The great advantage of this derivative definition is that for cases like this, where we have a derivative function $f'(x)$ that cannot be extended to include a point x_B because $\lim_{x\to x_B}f'(x)$ does not exist, the derivative $f'(x_B)$ also will fail to exist, just as we found the derivative $s'(0)$ failed to exist. For the point x_B, let us convert the line

$$\lim_{(x,h)\to(x_B,0)}\frac{f(x+h)-f(x)}{h}=f'(x_B)$$

into its requirement:

$$\frac{f(x+h)-f(x)}{h} \text{ defined and } |(x,h)-(x_B,0)| < \delta(\epsilon)$$

implies

$$\left|\frac{f(x+h)-f(x)}{h} - f'(x_B)\right| < \epsilon$$

In the interval $|x - x_B| < \delta(\epsilon)$, there will be points x such that $f'(x)$ exists. For any such x, we can let h of the quotient $(f(x+h) - f(x))/h$ approach 0 while x is held fixed, all the while staying in the neighborhood $|(x,h) - (x_B, 0)| < \delta(\epsilon)$. The limit we are taking is

$$\lim_{h\to 0} \frac{f(x+h)-f(x)}{h}$$

The limit equals $f'(x)$ because the derivative function $f'(x)$ is defined at x by the stronger limit of Final Definition 8.1. At the same time we have for all the values of h with x being held fixed, the inequality

$$\left|\frac{f(x+h)-f(x)}{h} - f'(x_B)\right| < \epsilon$$

In the interval $|x - x_B| < \delta(\epsilon)$ we obtain then the limit inequality

$$|f'(x) - f'(x_B)| \leq \epsilon$$

This implies $\lim_{x\to x_B} f'(x) = f'(x_B)$, so if we know that $\lim_{x\to x_B} f'(x)$ does not exist, a derivative at x_B is not possible. Thus, for the case of $s(x)$ with the derivative that could not be extended to $x = 0$, we did not need to attempt to compute $s'(0)$ because it was certain $s'(0)$ would not be defined.

An alternate form for the conventional derivative is

$$\lim_{x\to x_0} \frac{f(x)-f(x_0)}{x-x_0} = f'(x_0)$$

In this notation our derivative takes the form

$$\lim_{(x_1,x_2)\to(x_0,x_0)} \frac{f(x_1)-f(x_2)}{x_1-x_2} = f'(x_0)$$

Let $f(x)$ be a function defined and with a derivative on an interval I. The "divided difference" function $g(x,y)$ obtained from $f(x)$ by the definition

$$g(x,y) = \begin{cases} \frac{f(x)-f(y)}{x-y} & \text{if } x \neq y \\ f'(x) & \text{if } x = y \end{cases}$$

is a computable function defined for (x,y) in $x : I,\ y : I$, the region R. The function $(f(x) - f(y))/(x - y)$ is not defined in R along the line $y = x$, but for all x_0 in I we have $\lim_{(x,y)\to(x_0,x_0)} (f(x) - f(y))/(x - y) = f'(x_0)$, so according to Theorem 6.12 we have a computable function. We do not obtain this simple result if our derivative is defined by the Tentative Definition.

8.2 Rules of Differentiation

The stricter definition for the derivative that we use does not prevent obtaining the customary rules of differentiation, which we will need. These standard rules are listed in what follows. Here c denotes an arbitrary constant.

$$\begin{aligned}
(c)' &= 0 \\
(cf)' &= cf' \\
(x^n)' &= nx^{n-1} \\
(f+g)' &= f'+g' \\
(f-g)' &= f'-g' \\
(fg)' &= f'g+fg' \\
\left(\frac{f}{g}\right)' &= \frac{f'g-fg'}{g^2}
\end{aligned}$$

We consider in detail two of these rules, the rule for differentiating x^n and the product rule.

$$\begin{aligned}
(x^n)'_{x=x_0} &= \lim_{(x,h)\to(x_0,0)} \frac{(x+h)^n - x_n}{h} \\
&= \lim_{(x,h)\to(x_0,0)} \frac{\sum_{i=0}^{n} \binom{n}{i} x^{n-i}h^i - x^n}{h} = \lim_{(x,h)\to(x_0,0)} \frac{\sum_{i=1}^{n} \binom{n}{i} x^{n-i}h^i}{h} \\
&= \lim_{(x,h)\to(x_0,0)} \sum_{i=1}^{n} \binom{n}{i} x^{n-i}h^{i-1} = \binom{n}{1} x_0^{n-1} = nx_0^{n-1}
\end{aligned}$$

There is very little difference here between this proof and the conventional calculus proof. This is likewise the case for the product rule:

$$\begin{aligned}
(fg)'_{x=x_0} &= \lim_{(x,h)\to(x_0,0)} \frac{f(x+h)g(x+h) - f(x)g(x)}{h} \\
&= \lim_{(x,h)\to(x_0,0)} \frac{f(x+h)g(x+h) - f(x)g(x+h) + f(x)g(x+h) - f(x)g(x)}{h} \\
&= \lim_{(x,h)\to(x_0,0)} \frac{f(x+h) - f(x)}{h} g(x+h) + f(x)\frac{g(x+h) - g(x)}{h} \\
&= f'(x_0)g(x_0) + f(x_0)g'(x_0)
\end{aligned}$$

The proof of the quotient rule is also similar to the conventional proof.

The fundamental chain rule requires a careful statement:

THEOREM 8.1: Suppose $g(x)$ is defined on an interval I and $f(x)$ is defined on an interval I_1, and for every x in I we have $g(x)$ in I_1, so that the function $f(g(x))$ is defined on I. For any point x_0 in I, if $g'(x_0)$ exists and $f'(g(x_0))$ exists, then the derivative of $f(g(x))$ exists at $x = x_0$ and equals $f'(g(x_0))g'(x_0)$.

There is difficulty proving the chain rule if one tries to convert

$$\frac{\Delta f(g(x))}{\Delta x} \quad \text{into} \quad \frac{\Delta f(g(x))}{\Delta g(x)} \cdot \frac{\Delta g(x)}{\Delta x}$$

The problem here is that when $\Delta g(x)$ is zero, the quotient $\Delta f(g(x))/\Delta g(x)$ is not defined. A better procedure is to convert

$$\frac{\Delta f(g(x))}{\Delta x} \quad \text{into} \quad f'(g(x_0)) \cdot \frac{\Delta g(x)}{\Delta x} + \text{some remainder}$$

Then one proves a zero value for the limit of the remainder as Δx goes to zero.

Let us examine a conventional calculus treatment for obtaining the chain rule, for the case where x_0 and $g(x_0)$ are interior points of the respective intervals I and I_1. First a remainder function $r(x)$ is introduced, defined by the equation

$$r(x) = f(g(x)) - f(g(x_0)) - f'(g(x_0))(g(x) - g(x_0))$$

The derivative limit is then

$$\begin{aligned}\lim_{x \to x_0} \frac{f(g(x)) - f(g(x_0))}{x - x_0} &= \lim_{x \to x_0} \left[f'(g(x_0)) \frac{g(x) - g(x_0)}{x - x_0} + \frac{r(x)}{x - x_0} \right] \\ &= f'(g(x_0))g'(x_0) + \lim_{x \to x_0} \frac{r(x)}{x - x_0}\end{aligned}$$

The limit remaining is

$$\lim_{x \to x_0} \frac{r(x)}{x - x_0} = \lim_{x \to x_0} \frac{f(g(x)) - f(g(x_0)) - f'(g(x_0))(g(x) - g(x_0))}{x - x_0}$$

If $g(x) = g(x_0)$, the numerator on the right in the preceding equation is zero. If $g(x) \neq g(x_0)$, then we have

$$\begin{aligned}&\frac{f(g(x)) - f(g(x_0)) - f'(g(x_0))(g(x) - g(x_0))}{x - x_0} \\ &\quad = \left[\frac{f(g(x) - f(g(x_0))}{g(x) - g(x_0)} - f'(g(x_0)) \right] \frac{g(x) - g(x_0)}{x - x_0}\end{aligned}$$

Regardless of which case occurs, we have $\lim_{x\to x_0} r(x)/(x - x_0) = 0$, and an appropriate limit modulus can be constructed using the moduli for the two derivative limits $g'(x_0)$ and $f'(g(x_0))$, and the modulus for the limit $\lim_{x\to x_0} g(x) = g(x_0)$. In conventional calculus, this last limit is implied by the existence of the derivative $g'(x_0)$.

We can use a similar proof to establish the theorem, and because our limit definition is a relaxed form of the conventional limit definition, x_0 and $g(x_0)$ can be any points in their respective intervals I and I_1, not necessarily interior points. Our remainder function $r(x_1, x_2)$ is defined by the equation

$$r(x_1, x_2) = f(g(x_1)) - f(g(x_2)) - f'(g(x_0)(g(x_1) - g(x_2))$$

Our derivative limit is

$$\lim_{(x_1,x_2)\to(x_0,x_0)} \frac{f(g(x_1)) - f(g(x_2)}{x_1 - x_2} = \lim_{(x_1,x_2)\to(x_0,x_0)} \left[f'(g(x_0)) \frac{g(x_1) - g(x_2)}{x_1 - x_2} + \frac{r(x_1, x_2)}{x_1 - x_2} \right]$$
$$= f'(g(x_0))g'(x_0) + \lim_{(x_1,x_2)\to(x_0,x_0)} \frac{r(x_1, x_2)}{x_1 - x_2}$$

The limit remaining is

$$\lim_{(x_1,x_2)\to(x_0,x_0)} \frac{r(x_1, x_2)}{x_1 - x_2} = \lim_{(x_1,x_2)\to(x_0,x_0)} \frac{f(g(x_1)) - f(g(x_2)) - f'(g(x_0))(g(x_1) - g(x_2))}{x_1 - x_2}$$

If $g(x_1) = g(x_2)$, the numerator on the right in the preceding equation is zero. If $g(x_1) \neq g(x_2)$, then we have

$$\frac{f(g(x_1)) - f(g(x_2)) - f'(g(x_0))(g(x_1) - g(x_2))}{x_1 - x_2}$$
$$= \left[\frac{f(g(x_1)) - f(g(x_2))}{g(x_1) - g(x_2)} - f'(g(x_0)) \right] \frac{g(x_1) - g(x_2)}{x_1 - x_2}$$

Regardless of which case occurs, we have $\lim_{(x_1,x_2)\to(x_0,x_0)} r(x_1, x_2)/(x_1 - x_2) = 0$. Note that if we presume the case $g(x_1) \neq g(x_2)$ always holds, and construct a limit modulus using the other known moduli, the constructed modulus will serve for both cases.

8.3 A Computation Problem

When it is known that a function $f(x)$ has a derivative, but we have only the function without the derivative, sometimes it is difficult to accurately estimate the derivative at a point x_0 by forming an approximating quotient $(f(x_0 + h) - f(x_0))/h$. Here, without the derivative or the derivative limit modulus $\delta(\epsilon, x)$, we are uncertain what h value we should use. This difficulty can be made explicit as

NONSOLVABLE PROBLEM 8.1: Given a function $f(x)$ known to have a derivative at a point x_0, find $f'(x_0)$ to k or $k+1$ correct decimals.

Consider the function $f_a(x)$ with the parameter a, defined as follows:

$$f_a(x) = \begin{cases} a \sin \frac{x}{a} & \text{if } a \neq 0 \\ 0 & \text{if } a = 0 \end{cases} \tag{8.2}$$

We can view $f_a(x)$ as a function $g(x, a)$ of two variables, which is defined by the first line of Eq. (8.2) everywhere in the xa plane except on the line $a = 0$. Because for all x_0 we have $\lim_{(x,a)\to(x_0,0)} a \sin(x/a) = 0$, by Theorem 6.12 $g(x, a)$ is a computable function. Therefore, $f_a(x)$ values as accurate as we please can be obtained for any setting of the parameter a and any function argument x.

On the other hand, if we wanted to find the derivative of f_a at some point, there might be difficulty. If a is zero, $f_a'(x) = 0$, while if a is unequal to zero, $f_a'(x) = \cos(x/a)$. Thus, if $a \neq 0$, then $f_a'(0) = 1$, while if $a = 0$, then $f_a'(0) = 0$. There can be no computable procedure for finding the derivative of $f_a(x)$ at the point $x = 0$, for if we had such a procedure, from $f_a'(0)$ we would be able to determine whether or not the real number a was zero, and this contradicts Nonsolvable Problem 3.1.

8.4 The Mean Value Theorem

The Mean Value theorem given next takes a form like that of Theorem 6.8, in that the phrase "cannot fail to exist" is used instead of "exists" or "can be found." The nonsolvable problems given afterwards help justify this terminology.

THEOREM 8.2: If $f(x)$ is defined on $[a, b]$ with a derivative $f'(x)$ on (a, b), then in the interval (a, b) a point x_0 cannot fail to exist such that $f'(x_0) = (f(b) - f(a))/(b - a)$.

In conventional calculus it is customary to prove this theorem by first proving a result known as Rolle's lemma or Rolle's theorem, a special case of the Mean Value theorem, where $f(a) = f(b) = 0$. The proof of Rolle's lemma is obtained by showing that a continuous function on $[a, b]$ satisfying these conditions must assume a maximum or a minimum value at a point in (a, b), and at such a point the derivative is zero. However, Theorem 7.4 has shown that in computable calculus, it is possible for a function uniformly continuous on $[a, b]$ not to assume a maximum or a minimum value. Recall that the function

$s(x)$ of Theorem 7.4 has a least upper bound, but never attains that value. Therefore, our proof of the Mean Value Theorem is structured differently, without the usual lemma of conventional calculus.

We assume there is no point x_0 in (a, b) with a derivative that matches the number $c = (f(b) - f(a))/(b - a)$, and show that this assumption leads to a contradiction. We define a monotone increasing sequence a_n and a monotone decreasing sequence b_n, with all sequence terms in (a, b), such that for all n we have $0 < b_n - a_n < (\frac{1}{2})^{n-1} \cdot (b_1 - a_1)$. This implies that both sequences a_n and b_n converge to a common limit L in (a, b). We show that $f'(L) = c$, giving us the contradiction.

To define a_1 and b_1, let m_0 be the midpoint of the interval $[a, b]$. Because $f(m_0) \neq c$, when we obtain increasingly accurate approximation intervals for $f'(m_0)$ and c, eventually we must obtain intervals that are disjoint, and then we find either $f'(m_0) > c$ or $f'(m_0) < c$. For clarity we suppose $f'(m_0) > c$. The other case $f'(m_0) < c$ is treated analogously. Let $\delta(\epsilon, x)$ be the limit modulus for obtaining the derivative $f'(x)$ from $f(x)$. We choose ϵ_0 less than $f'(m_0) - c$ and take δ_0 equal to $\frac{1}{2}\delta(\epsilon_0, m_0)$. Then for any two distinct numbers x_1, x_2 in the *closed* interval $m_0 \pm \delta_0$, we have $(f(x_1) - f(x_2))/(x_1 - x_2) > c$. Using the same reasoning as was used in the discussion of the derivative definition, for any number x in $m_0 \pm \delta_0$, we have $f'(x) > c$. In particular, $(f(m_0 + \delta_0) - f(m_0 - \delta_0))/2\delta_0 > c$, $f'(m_0 - \delta_0) > c$, and $f'(m_0 + \delta_0) > c$.

The $f(x)$ slope calculated for the interval $[a, b]$ equals c and is the weighted average of the $f(x)$ slopes computed over intervals $[a, m_0 - \delta_0]$, $[m_0 - \delta_0, m_0 + \delta_0]$, and $[m_0 + \delta_0, b]$, the weights being $(m_0 - \delta_0 - a)/(b - a)$, $2\delta_0/(b - a)$, and $(b - m_0 - \delta_0)/(b - a)$, respectively, these weights summing to 1. That is, we have

$$\begin{aligned}\frac{f(b) - f(a)}{b - a} = c = {} & \frac{m_0 - \delta_0 - a}{b - a} \cdot \frac{f(m_0 - \delta_0) - f(a)}{m_0 - \delta_0 - a} \\ & + \frac{2\delta_0}{b - a} \cdot \frac{f(m_0 + \delta_0) - f(m_0 - \delta_0)}{2\delta_0} \\ & + \frac{b - m_0 - \delta_0}{b - a} \cdot \frac{f(b) - f(m_0 + \delta_0)}{b - m_0 - \delta_0}\end{aligned}$$

The computed slope on the middle interval is greater than c, and this implies that for at least one of the two intervals $[a, m_0 - \delta_0]$, $[m_0 + \delta_0, b]$, the computed $f(x)$ slope is $< c$. When we compute $f(x)$ slopes on these two intervals to sufficient accuracy, we can determine one of these slopes to be $< c$. If we determine the slope on $[a, m_0 - \delta_0]$ is $< c$, then we take $b_1 = m_0 - \delta_0$, and take a_1 to be a point to the right of a but close enough so that the $f(x)$ slope on $[a_1, b_1]$ also is $< c$. If we determine that the slope on $[m_0 + \delta_0, b]$ is $< c$, then we take $a_1 = m_0 + \delta_0$, and take b_1 to be a point to the left of b but close enough so that the $f(x)$ slope on $[a_1, b_1]$ also is $< c$.

After a_1 and b_1 are determined, the sequence terms a_n and b_n for $n > 1$ are found recursively. We suppose that for all n the $f(x)$ slope on $[a_n, b_n]$ is $< c$, and that at one of the endpoints a_n, b_n, the derivative of $f(x)$ is $> c$. These conditions are satisfied for $n = 1$. The general procedure is similar to the procedure for the case $n = 1$. We take m_n equal to the midpoint $(a_n + b_n)/2$ of the interval $[a_n, b_n]$. We obtain ever more accurate approximation intervals for $f'(m_n)$ and for c, and eventually we obtain intervals that are disjoint. We assume that for all n we always find $f'(m_n) > c$, because if we ever find the opposite, say $f'(m_N) < c$, then the interval with endpoints m_0 and m_N is such that $f'(m_0) > c$ and $f'(m_N) < c$. According to Theorem 6.8, a point x_0 between these endpoints satisfying the equation $f'(x_0) = c$ cannot fail to exist, and so we have an immediate contradiction.

Again we can find a positive number δ_n such that $f'(x) > c$ for x in $m_n \pm \delta_n$, and such that the $f(x)$ slope computed over the interval $m_n \pm \delta_n$ is $> c$. The $f(x)$ slope computed over $[a_n, b_n]$ is $< c$ and is the weighted average of $f(x)$ slopes computed over the intervals $[a_n, m_n - \delta_n]$, $[m_n - \delta_n, m_n + \delta_n]$, and $[m_n + \delta_n, b_n]$. Again we determine either the interval $[a_n, m_n - \delta_n]$ or the interval $[m_n + \delta_n, b_n]$ to have a computed $f(x)$ slope $< c$. In the first case we take $a_{n+1} = a_n$ and $b_{n+1} = m_n - \delta_n$; in the other case we take $a_{n+1} = m_n + \delta_n$ and $b_{n+1} = b_n$.

Now let the common limit of the two sequences a_n and b_n be the number L in (a, b). We have $(f(b_n) - f(a_n))/(b_n - a_n) < c$, and taking the limit with respect to n, we obtain $f'(L) \leq c$. On the other hand, any neighborhood of L contains sequence terms a_n or b_n at which the derivative $f'(x)$ is $> c$, and this implies $f'(L) \geq c$. The two inequalities together imply $f'(L) = c$ and we have a contradiction.

COROLLARY 1: If $f(x)$ is defined on $[a, b]$ and has an identically zero derivative on (a, b), then $f(x)$ is constant on $[a, b]$.

If x is any number in $(a, b]$, then by the Mean Value theorem, we have $f(x) - f(a) = f'(x_0)(x - a) = 0(x - a) = 0$. Therefore, $f(x)$ equals $f(a)$ for all x in $(a, b]$.

COROLLARY 2: If $f(x)$ is defined on $[a, b]$ and has a bounded derivative on (a, b), then $f(x)$ is uniformly continuous on $[a, b]$.

We have $|f'(x)| \leq M$. If x_1 and x_2 are any two distinct points in $[a, b]$, then applying the theorem, a point x_0 in the interior of the closed interval having endpoints x_1 and x_2 cannot fail to exist such that

$$\frac{f(x_1) - f(x_2)}{x_1 - x_2} = f'(x_0)$$

Taking absolute values, we obtain

$$|f(x_1) - f(x_2)| = |f'(x_0)||x_1 - x_2| \leq M|x_1 - x_2|$$

The modulus $\delta(\epsilon)$ required for uniform continuity may be taken equal to ϵ/M.

Here is a relevant nonsolvable problem.

NONSOLVABLE PROBLEM 8.2: Given a function $f(x)$ defined on $[a, b]$ with a derivative $f'(x)$ defined on (a, b), find to k or $k+1$ correct decimals a number x_0 in (a, b) with $f'(x_0) = (f(b) - f(a))/(b - a)$.

Here we make use of the polynomial $p_1(x) = x(x-1)^2$ with derivative $(x-1)(3x-1)$, and the polynomial $p_2(x) = x^2(x-1)$ with derivative $x(3x-2)$. Both polynomials are zero at $x = 0$ and at $x = 1$. If we take $[a, b]$ to be $[0, 1]$, the point x_0 equals $\frac{1}{3}$ for the function $p_1(x)$ and equals $\frac{2}{3}$ for the function $p_2(x)$. We define a function $f_a(x)$ with the parameter a by the equation

$$f_a(x) = \begin{cases} a \cdot p_1(x) & \text{if } a > 0 \\ 0 & \text{if } a = 0 \\ a \cdot p_2(x) & \text{if } a < 0 \end{cases} \tag{8.3}$$

This is a computable function as may be seen by the following reasoning. Let $h(x, a)$ be a function that is defined by the first and third lines of Eq. (8.3). The function $h(x, a)$ is defined everywhere in the xa plane except on the line $a = 0$. For all x_0 we have $\lim_{(x,a)\to(x_0,0)} h(x, a) = 0$, so by Theorem 6.12 this function can be extended to a function $g(x, a)$ that is defined by the right-hand side of Eq. (8.3). This implies that $f_a(x)$ is defined for all x and for any value of the parameter a.

The derivative $f_a'(x)$ is given by the equation

$$f_a'(x) = \begin{cases} a \cdot p_1'(x) & \text{if } a > 0 \\ 0 & \text{if } a = 0 \\ a \cdot p_2'(x) & \text{if } a < 0 \end{cases}$$

This is a computable function by the same reasoning as given for $f_a(x)$. If we could always obtain for the function $f_a(x)$ over the interval $[0, 1]$ a correct decimal value for the point x_0 of the problem, then if the supplied x_0 value was $< \frac{1}{2}$, we would be certain that $a \geq 0$, and if the supplied x_0 value was $> \frac{1}{2}$, we would be certain that $0 \geq a$. The supplied correct decimal value defines an interval which may contain the $\frac{1}{2}$ point, and in this case we can be certain that

$a = 0$. This contradicts Nonsolvable Problem 3.12, so the imagined procedure is not possible.

We obtain a solvable problem by reducing our requirement.

SOLVABLE PROBLEM 8.3: Given a function $f(x)$ defined on $[a, b]$ with a derivative $f'(x)$ defined on (a, b), find to k or $k + 1$ correct decimals a number x_0 in (a, b) such that $f'(x_0) - (f(b) - f(a))/(b - a) = 0.00\langle k \text{ zeros}\rangle 00{\sim}$.

Note that the problem implies that $|f'(x_0) - (f(b) - f(a))/(b - a)| \leq \frac{1}{2}10^{-k}$. Here we can follow a procedure much like that described in the Mean Value theorem, testing midpoints m_n of progressively smaller intervals $[a_n, b_n]$, for $n = 0, 1, 2, \ldots$, with a_0 taken as a and b_0 taken as b. For each n we obtain $D = k + 1$ interval approximations for $f'(m_n)$ and for $c = (f(b) - f(a))/(b - a)$. If for any n we obtain overlapping intervals for c and for $f'(m_n)$, then $|f'(m_n) - c| \leq 4 \cdot 10^{-(k+1)}$ and we satisfy this problem by taking m_n as x_0 and supplying this number to k or $k + 1$ correct decimal places. For $n = 0$ and the case where the intervals for $f'(m_0)$ and c are disjoint, we choose a_1 and b_1 by the method described previously.

If we ever find for some n larger than 0 that $f'(m_n)$ and $f'(m_0)$ are on opposite sides of c, then we apply the procedure of Solvable Problem 6.2 to the function $f'(x)$ to obtain a point x_0 between m_0 and m_n that satisfies the problem. Otherwise we continue the process until for some n, say N, the length of $[a_N, b_N]$ first becomes $< 10^{-(k+2)}$. According to the theorem, a number x_1 in $[a_N, b_N]$ cannot fail to exist such that $f'(x_1) = (f(b_N) - f(a_N))/(b_N - a_N)$. The derivatives at x_1 and either at a_N or at b_N are on opposite sides of c. According to Theorem 6.8 we can conclude that a number x_0 in $[a_N, b_N]$ cannot fail to exist such that $f'(x_0) = c$. Now we compute a $D = k + 2$ interval for a_N (or for b_N), with the error bound of 1 unit in the last decimal place increased to 2 units to reflect the size of $[a_N, b_N]$. This approximation can be rounded to k places or $k + 1$ places to give the appropriate x_0 decimal approximation. (The second paragraph after Solvable Problem 3.8 is relevant here.)

CHAPTER 9

The Riemann Integral

9.1 Riemann Sums

The theory of the Riemann integral is easily accommodated in computable calculus. The first step is to define a partition of a finite interval and related concepts.

DEFINITION 9.1: *A* partition *of the interval* $[a, b]$ *is given by a positive integer* N *and a finite sequence* x_i *defined for* $0 \leq i \leq N$*, such that* $x_0 = a$*,* $x_N = b$*, and* $x_{i-1} < x_i$ *for* $1 \leq i \leq N$*. The partition* norm*, denoted by the symbol* $||x_i||$*, is the number*

$$||x_i|| = \max_{1 \leq i \leq N} (x_i - x_{i-1})$$

A selection sequence *for a partition* N*,* x_i *is a finite sequence* $\widehat{x}_i$*, satisfying the relations* $x_{i-1} \leq \widehat{x}_i \leq x_i$ *for* $1 \leq i \leq N$*. The partition is said to have* right selection *(*left selection*) if the term* $\widehat{x}_i$ *is always the right (left) endpoint of the interval* $[x_{i-1}, x_i]$*.*

Given a function $f(x)$ defined on $[a, b]$, there is a corresponding Riemann sum $\sum_{i=1}^{N} f(\widehat{x}_i)(x_i - x_{i-1})$ for every partition N, x_i with accompanying selection sequence $\widehat{x}_i$. The Riemann integral is defined in terms of these sums.

DEFINITION 9.2: *Suppose the function* $f(x)$ *is defined on an interval* $[a, b]$*. There is a certain number, the* integral of $f(x)$ over $[a, b]$*, denoted by the symbol*

$\int_a^b f(x)\,dx$, *if there is a positive-valued semifunction* $\delta(\epsilon)$ *of the positive variable* ϵ *such that for any partition* N, x_i *with selection sequence* $\widehat{x}_i$,

$$||x_i|| < \delta(\epsilon) \quad \textit{implies} \quad \left|\sum_{i=1}^{N} f(\widehat{x}_i)(x_i - x_{i-1}) - \int_a^b f(x)\,dx\right| < \epsilon \tag{9.1}$$

The preceding definition may be expressed in symbolic form as:

$$\lim_{||x_i|| \to 0} \sum_{i=1}^{N} f(\widehat{x}_i)(x_i - x_{i-1}) = \int_a^b f(x)\,dx$$

A function $f(x)$ is *integrable* on $[a, b]$ if it has an integral on this interval. And when we write some integral $\int_a^b g(x)\,dx$, the function $g(x)$ is called the *integrand*.

9.2 The Integration of Uniformly Continuous Functions

In conventional calculus, a standard result is that a function $f(x)$ defined and continuous on $[a, b]$ is integrable on the interval. In computable calculus, a function defined on $[a, b]$ is continuous there, but nevertheless may not be integrable on the interval. For instance, the function $t(x)$ used as the example of Theorem 6.9, is defined and necessarily continuous on $[0, 1]$, but this function is unbounded on $[0, 1]$. Recall that $t(b_n) = n$, while at the neighboring point b'_n equal to $b_n + 10^{-(n+1)}$, we have $t(b'_n) = 0$. The function $t(x)$ is not integrable on $[0, 1]$, because for any n greater than 10, we can choose a partition with norm equal to $1/n$, making sure to have the points b_n and $b_n + 1/n$ as two successive division points x_{k-1} and x_k. The Riemann sum obtained with left selection is larger by 1 than the same Riemann sum except that the single point $\widehat{x}_k$ is shifted to b'_n. Thus for the function $t(x)$, no matter how small the norm of a partition is required to be, we can never get the variation of Riemann sums < 1.

However, any function uniformly continuous on an interval $[a, b]$ is integrable there, and the proof of this result is similar to the proof of the conventional calculus result mentioned in the preceding paragraph.

THEOREM 9.1: A function $f(x)$ uniformly continuous on $[a, b]$ is integrable on this interval.

Our first step is to obtain a general result linking pairs of partitions. If N_1, u_i and N_2, v_i are two partitions of $[a, b]$, there is another partition N, x_i that uses all the distinct division points of u_i and v_i. We have $x_0 = a$ and $x_N = b$, as usual, and as i increases from 1 to $N-1$, the term x_i equals all the u_i and v_i terms that lie in (a, b).

Notice here the difficulty in actually constructing the partition N, x_i, because this requires deciding whether a division point u_i is less than, equal to, or greater than a division point v_i. We can use here the principle of finite choice (see Section 6.2), since clearly a certain finite number of alternative partitions N, x_i could be constructed, with one of them having the properties proposed.

Riemann sums for either partition N_1, u_i or N_2, v_i can be expressed in terms of the partition N, x_i, because a Riemann sum multiplier $(u_i - u_{i-1})$ or $(v_i - v_{i-1})$ equals a multiplier $(x_i - x_{i-1})$ or a sum of such multipliers.

$$\sum_{i=1}^{N_1} f(\widehat{u}_i)(u_i - u_{i-1}) = \sum_{i=1}^{N} f(\widehat{u}_{j_i})(x_i - x_{i-1})$$

$$\sum_{i=1}^{N_2} f(\widehat{v}_i)(v_i - v_{i-1}) = \sum_{i=1}^{N} f(\widehat{u}_{k_i})(x_i - x_{i-1})$$

Here j_i is an integer-valued sequence defined for $1 \leq i \leq N$ that gives the appropriate $\widehat{u}_i$ index to use with the multiplier $(x_i - x_{i-1})$, and the sequence k_i plays the same role for $\widehat{v}_i$. Thus, if $(u_1 - u_0) = (x_2 - x_1) + (x_1 - x_0)$, then $j_1 = 1$ and $j_2 = 1$.

Because $f(x)$ is uniformly continuous on $[a, b]$, there is a semifunction $\delta_1(\epsilon)$ such that for any two points x_1, x_2 in $[a, b]$,

$$|x_1 - x_2| < \delta_1(\epsilon) \quad \text{implies} \quad |f(x_1) - f(x_2)| < \epsilon$$

If both partitions N_1, u_i and N_2, v_i have norms less than $\frac{1}{2}\delta_1(\epsilon)$, then the difference between their Riemann sums is less than $\epsilon(b-a)$:

$$\begin{aligned}\left|\sum_{i=1}^{N_1} f(\widehat{u}_i)(u_i - u_{i-1}) - \sum_{i=1}^{N_2} f(\widehat{v}_i)(v_i - v_{i-1})\right| &= \left|\sum_{i=1}^{N}\Big(f(\widehat{u}_{j_i}) - f(\widehat{v}_{k_i})\Big)(x_i - x_{i-1})\right| \\ &\leq \sum_{i=1}^{N}\left|f(\widehat{u}_{j_i}) - f(\widehat{v}_{k_i})\right|(x_i - x_{i-1}) \\ &< \sum_{i=1}^{N}\epsilon(x_i - x_{i-1}) = \epsilon(b-a)\end{aligned}$$

To obtain the last displayed line, we note that $\widehat{u}_{j_i}$ and $\widehat{v}_{k_i}$ are selection terms that lie within certain intervals $[u_{j_i-1}, u_{j_i}]$ and $[v_{k_i-1}, v_{k_i}]$ that intersect because they both contain $[x_{i-1}, x_i]$. Therefore, the distance between $\widehat{u}_{j_i}$ and $\widehat{v}_{k_i}$ is not greater than the sum of the two norms $||u_i||$ and $||v_i||$, and this sum is $< \delta_1(\epsilon)$.

We now use this general result to express the integral as the limit of a certain Cauchy sequence. For any positive integer n, let the partition n, $x_i^{(n)}$ with norm $(b-a)/n$ be defined by taking $x_i^{(n)}$ equal to $a+i(b-a)/n$. We use right selection and define a sequence d_n by the equation

$$d_n = \sum_{i=1}^{n} f(x_i^{(n)})(x_i^{(n)} - x_{i-1}^{(n)}) = \sum_{i=1}^{n} f(x_i^{(n)})\frac{b-a}{n}$$

To prove the sequence d_n is Cauchy, we must construct a semifunction $N_C(\epsilon)$ such that for $n_1, n_2 > N_C(\epsilon)$ we have $|d_{n_1} - d_{n_2}| < \epsilon$. From our general result about pairs of partitions, it follows that when the norms $(b-a)/n_1$ and $(b-a)/n_2$ are both $< \frac{1}{2}\delta_1(\epsilon)$, the absolute value of the difference between d_{n_1} and d_{n_2} is $< \epsilon(b-a)$. That is, $(b-a)/n_1, (b-a)/n_2 < \frac{1}{2}\delta_1(\epsilon)$ implies $|d_{n_1} - d_{n_2}| < \epsilon(b-a)$. This is equivalent to $n_1, n_2 > (b-a)/\frac{1}{2}\delta_1(\epsilon)$ implies $|d_{n_1} - d_{n_2}| < \epsilon(b-a)$, so the semifunction $N_C(\epsilon)$ may be taken equal to $(b-a)/\frac{1}{2}\delta_1(\epsilon/(b-a))$.

The next step is to show that we can find a semifunction $\delta(\epsilon)$ for which Eq. (9.1) is true, with $\int_a^b f(x)\,dx$ taken as the limit L of the sequence d_n. For any partition N, x_i such that $||x_i|| < \frac{1}{2}\delta_1(\epsilon)$, for any accompanying selection sequence $\widehat{x}_n$, and for any integer n, we have the following relations:

$$\begin{aligned}\left|\sum_{i=1}^{N} f(\widehat{x}_i)(x_i - x_{i-1}) - \int_a^b f(x)\,dx\right| &= \left|\sum_{i=1}^{N} f(\widehat{x}_i)(x_i - x_{i-1}) - d_n + d_n \right.\\ &\qquad \left. - \int_a^b f(x)\,dx\right| \\ &\le \left|\sum_{i=1}^{N} f(\widehat{x}_i)(x_i - x_{i-1}) - d_n\right| + |d_n - L|\end{aligned}$$

If we choose n large enough so that the d_n norm $(b-a)/n$ is $< \frac{1}{2}\delta_1(\epsilon)$, and also large enough so that the distance between d_n and L is $< \epsilon(b-a)$, we obtain

$$\left|\sum_{i=1}^{N} f(\widehat{x}_i)(x_i - x_{i-1}) - \int_a^b f(x)\,dx\right| < \epsilon(b-a) + \epsilon(b-a) = 2\epsilon(b-a)$$

This shows that the function $\delta(\epsilon)$ may be taken as $\frac{1}{2}\delta_1(\epsilon/2(b-a))$.

9.3 Properties of Integrals

The properties of integrals that are used in conventional calculus are valid in computable calculus for uniformly continuous integrands.

THEOREM 9.2: Let the functions $f(x)$, $f_1(x)$, and $f_2(x)$ be uniformly continuous on the interval $[a, b]$, and let c, c_1, and c_2 be arbitrary constants. Then

(a) $\int_a^b c\,dx = c(b-a)$

(b) $\int_a^b (c_1 f_1(x) + c_2 f_2(x))\,dx = c_1 \int_a^b f_1(x)\,dx + c_2 \int_a^b f_2(x)\,dx$

(c) If for all x in $[a, b]$ we have $f_1(x) \le f_2(x)$, then $\int_a^b f_1(x)\,dx \le \int_a^b f_2(x)\,dx$.

(d) $\left|\int_a^b f(x)\,dx\right| \le \int_a^b |f(x)|\,dx$

(e) For any number x_0 in (a, b), $\int_a^b f(x)\,dx = \int_a^{x_0} f(x)\,dx + \int_{x_0}^b f(x)\,dx$.

PROOF: The proofs are obtained by using appropriate Cauchy sequences d_n, as defined in the preceding section, which converge to $\int_a^b f(x)\,dx$ when the function $f(x)$ is uniformly continuous on $[a, b]$. Let us use the notation $d_n^{(f)}$ for the $f(x)$ sequence. For the first two cases, we obtain

$$\text{(a)} \quad \int_a^b c\,dx = \lim_{n\to\infty} d_n^{(c)} = \lim_{n\to\infty} \sum_{i=1}^{n} c\frac{b-a}{n} = \lim_{n\to\infty} c(b-a) = c(b-a)$$

$$\text{(b)} \quad \int_a^b (c_1 f_1(x) + c_2 f_2(x))dx = \lim_{n\to\infty} d_n^{(c_1 f_1 + c_2 f_2)} = \lim_{n\to\infty} c_1 d_n^{(f_1)} + c_2 d_n^{(f_2)}$$

$$= c_1 \int_a^b f_1(x)\,dx + c_2 \int_a^b f_2(x)\,dx$$

The hypothesis of (c) implies $d_n^{(f_1)} \le d_n^{(f_2)}$. Consequently,

$$\int_a^b f_1(x)\,dx = \lim_{n\to\infty} d_n^{(f_1)} \le \lim_{n\to\infty} d_n^{(f_2)} = \int_a^b f_1(x)\,dx$$

The relation (d) is obtained by applying (c) to the inequalities

$$-|f(x)| \le f(x) \le |f(x)|$$

to obtain

$$-\int_a^b |f(x)|\,dx \le \int_a^b f(x)\,dx \le \int_a^b |f(x)|\,dx$$

which implies

$$\left|\int_a^b f(x)\,dx\right| \le \int_a^b |f(x)|\,dx$$

And finally the relation (e) is obtained by defining sequences d_n' and d_n'' for $f(x)$ on $[a, x_0]$ and $[x_0, b]$, respectively. Define the sequence c_n by setting $c_n = d_n' + d_n''$. The term c_n equals a Riemann sum of $f(x)$ on $[a, b]$ with norm equal to the maximum of the norms for d_n' and for d_n''. The relation (9.1) implies that $\lim_{n\to\infty} c_n$ equals $\int_a^b f(x)\,dx$, so we obtain

$$\int_a^b f(x)\,dx = \lim_{n\to\infty} (d_n' + d_n'') = \int_a^{x_0} f(x)\,dx + \int_{x_0}^b f(x)\,dx$$

This concludes the proof of the theorem. ■

In order to have relation (e) be valid for x_0 equal to a, equal to b, or equal to some point outside $[a, b]$, it is customary to extend the interpretation of $\int_a^b f(x)\,dx$ as follows ($a < b$ is presumed):

$$\int_b^a f(x)\,dx = -\int_a^b f(x)\,dx$$

$$\int_a^a f(x)\,dx = 0$$

Then if $f(x)$ is uniformly continuous on an interval $[a_1, b_1]$ that contains $[a, b]$, relation (e) holds for any number x_0 in $[a_1, b_1]$.

9.4 Defining a Function by Means of an Integral

If $f(x)$ is uniformly continuous on $[a, b]$, then by the results of the previous section, for each number x in $[a, b]$ there is a corresponding number $\int_a^x f(t)\,dt$. However, we have not shown that $\int_a^x f(t)\,dt$ satisfies our Definition 4.2 for functions of one variable. We do this next.

THEOREM 9.3: If $f(x)$ is uniformly continuous on $[a, b]$, there is a function $g(x)$, also uniformly continuous on $[a, b]$, such that

(a) $g(x) = \int_a^x f(t)\,dt$ for x in $[a, b]$.

(b) $g'(x)$ exists and equals $f(x)$ for x in $[a, b]$.

Define the sequence of functions $g_n(x)$ on the interval $[a, b]$ by the equation

$$g_n(x) = \frac{x-a}{n} \sum_{i=1}^{n} f\left(a + i\frac{x-a}{n}\right)$$

If x is in $(a, b]$, then the right-hand side of the preceding equation is a Riemann sum on $[a, x]$ for the partition $N = n$, $x_i = a + i(x-a)/n$, and with right selection. It was shown in the proof of Theorem 9.1, that if $\delta_1(\epsilon)$ is a uniform continuity modulus for $f(x)$ on $[a, b]$, then for any two partitions of $[a, b]$ with norms $< \frac{1}{2}\delta_1(\epsilon)$, regardless of how a selection sequence for each partition is chosen, the absolute value of the difference between the two Riemann sums is $< \epsilon(b-a)$. The uniform continuity modulus $\delta_1(\epsilon)$ for $[a, b]$ also serves as a uniform continuity modulus for the interval $[a, x]$. Consequently,

$$\frac{b-a}{n_1}, \frac{b-a}{n_2} < \frac{1}{2}\delta_1(\epsilon) \quad \text{implies} \quad |g_{n_1}(x) - g_{n_2}(x)| < \epsilon(x-a)$$

This shows that the sequence of functions $g_n(x)$ is uniformly Cauchy on $[a, b]$. ($N_C(\epsilon)$ may be taken as $(b-a)/\frac{1}{2}\delta_1(\epsilon/(b-a))$.) According to Theorem 6.10, the sequence of functions $g_n(x)$ converges uniformly to a limit function $g(x)$, and equation (a) for this limit function is implied by the definition of $g_n(x)$. Also, according to Theorem 7.1, each function $g_n(x)$ is uniformly continuous on $[a, b]$, and then Theorem 7.2 implies that the limit function $g(x)$ is also uniformly continuous on $[a, b]$.

Finally, to obtain result (b), we write

$$\begin{aligned}\frac{g(x_1) - g(x_2)}{x_1 - x_2} - f(x_0) &= \frac{1}{x_1 - x_2}\int_{x_2}^{x_1} f(t)\,dt - f(x_0) \\ &= \frac{1}{x_1 - x_2}\int_{x_2}^{x_1} f(t)\,dt - \frac{1}{x_1 - x_2}\int_{x_2}^{x_1} f(x_0)\,dt \\ &= \frac{1}{x_1 - x_2}\int_{x_2}^{x_1} (f(t) - f(x_0))\,dt\end{aligned}$$

Taking absolute values, we obtain

$$\begin{aligned}\left|\frac{g(x_1) - g(x_2)}{x_1 - x_2} - f(x_0)\right| &= \frac{1}{|x_1 - x_2|}\left|\int_{x_2}^{x_1} (f(t) - f(x_0))\,dt\right| \\ &\leq \frac{1}{|x_1 - x_2|}\int_{\min(x_1,x_2)}^{\max(x_1,x_2)} |f(t) - f(x_0)|\,dt\end{aligned}$$

If $|x_1 - x_0|, |x_2 - x_0| < \delta_1(\epsilon)$, then we have

$$\begin{aligned}\left|\frac{g(x_1) - g(x_2)}{x_1 - x_2} - f(x_0)\right| &< \frac{1}{|x_1 - x_2|}\int_{\min(x_1,x_2)}^{\max(x_1,x_2)} \epsilon\,dx \\ &= \frac{1}{|x_1 - x_2|}\epsilon|x_1 - x_2| = \epsilon\end{aligned}$$

This shows that $g(x)$ has the required derivative. Notice that $g'(x)$ exists in $[a, b]$ and not just in (a, b) as would be the case with conventional calculus. This difference is brought about by the changed derivative definition and the changed limit definition.

9.5 A Mean Value Theorem for Integrals

The theorem listed in what follows becomes the conventional calculus result if "uniformly continuous" is changed to "continuous," and "cannot fail to exist" is changed to "exists."

THEOREM 9.4: If $f(x)$ is uniformly continuous on $[a, b]$, then a number x_0 in (a, b) cannot fail to exist such that

$$\int_a^b f(x)\,dx = f(x_0)(b-a)$$

For the function $g(x)$, which equals

$$g(x) = \int_a^x f(t)\,dt$$

we have $g(a) = 0$, $g(b) = \int_a^b f(x)\,dx$, and $g'(x)$ is defined and equals $f(x)$ for x in $[a, b]$. When we apply the Mean Value theorem, we find that there cannot fail to exist a point x_0 in (a, b) with $g'(x_0)$ equal to $(\int_a^b f(x)\,dx - 0)/(b-a)$, and this implies the result we are trying to prove.

As usual, to justify using the expression "cannot fail to exist," we list a relevant nonsolvable problem:

NONSOLVABLE PROBLEM 9.1: Given a uniformly continuous function $f(x)$ defined on an interval $[a, b]$, find, to k or $k+1$ correct decimals, a number x_0 such that

$$\int_a^b f(x)\,dx = f(x_0)(b-a)$$

Because $f(x)$ is uniformly continuous on $[a, b]$, the function $g(x) = \int_a^x f(t)\,dt$ is defined on the interval $[a, b]$ and has a derivative everywhere there. Thus our problem is a special case of Nonsolvable Problem 8.2, and the method used there may be used again to show this problem is nonsolvable.

The problem can be made solvable in this manner:

SOLVABLE PROBLEM 9.2: Given a uniformly continuous function $f(x)$ defined in an interval $[a, b]$, find to k or $k+1$ correct decimals, a number x_0 such that

$$\int_a^b f(x)\,dx - f(x_0)(b-a) = 0.00\langle k \text{ zeros}\rangle 00\sim$$

Here we may use the procedure of Solvable Problems 8.2.

CHAPTER 10

Functions of Two Variables

In this chapter we take the domain region R of a function $f(x, y)$ to be given by one or more generalized intervals $I^{(2)}$, discussed in Section 4.9. For example, an unbounded domain R could take the form $x: [a, \infty), y: (-\infty, g(x))$.

10.1 Partial Derivatives

In conventional calculus the partial derivative of $f(x, y)$ with respect to x at a point (x_0, y_0) is taken to be the ordinary derivative of the function $f(x, y_0)$, that is, the derivative of the function $g(x) = f(x, y_0)$ obtained by holding y fixed at y_0. If we follow this plan but use our version of the ordinary derivative, then we have

TENTATIVE DEFINITION 10.1: *The partial derivative with respect to x of a function $f(x, y)$ at the point (x_0, y_0) is a real number, to be denoted by the symbol $\partial f/\partial x(x_0, y_0)$, provided the limit exists:*

$$\lim_{(x,h)\to(x_0,0)} \frac{f(x+h, y_0) - f(x, y_0)}{h} = \frac{\partial f}{\partial x}(x_0, y_0)$$

Suppose the function $f(x, y)$ is defined on a region R. The partial derivative of $f(x, y)$ with respect to x is a function $\partial f/\partial x(x, y)$ defined on R, provided the given limit exists for any point (x_0, y_0) in R. The partial derivative with respect to y is defined analogously.

This definition, even though it incorporates the stricter derivative limit of Chapter 8, is still too lenient. All the standard rules for partial derivatives would be obtained, but there is a fundamental difficulty. Consider the function $g(x, y)$ defined here:

$$g(x, y) = \begin{cases} xy \sin(1/\sqrt{x^2 + y^2}) & \text{if } (x, y) \neq (0, 0) \\ 0 & \text{if } (x, y) = (0, 0) \end{cases}$$

The function $xy \sin(1/\sqrt{x^2 + y^2})$ is defined for all (x, y) except for $(x, y) = (0, 0)$, and $\lim_{(x,y)\to(0,0)} xy \sin(1/\sqrt{x^2 + y^2}) = 0$, so according to Theorem 6.11, the function $g(x, y)$ is defined everywhere. The partial derivative function $\partial g/\partial x(x, y)$, defined everywhere except at $(x, y) = (0, 0)$, may be obtained by applying the standard rules of differentiation to $xy \sin(1/\sqrt{x^2 + y^2})$:

$$\frac{\partial g}{\partial x}(x, y) = y \sin(1/\sqrt{x^2 + y^2}) - [x^2 y/(x^2 + y^2)^{3/2}] \cos(1/\sqrt{x^2 + y^2})$$

For $\partial g/\partial x(0, 0)$ we obtain the value 0 because

$$\lim_{(x,h)\to(0,0)} \frac{g(x + h, 0) - g(x, 0)}{h} = \lim_{(x,h)\to(0,0)} = \frac{0 - 0}{h} = 0$$

The zero value for $\partial g/\partial x(0, 0)$ cannot be adjoined to the partial derivative function because this creates a function discontinuous at (0, 0). To see this, set y equal to x in the function $\partial g/\partial x(x, y)$ obtained in the proceeding and then take the limit as x approaches 0. We find:

$$\begin{aligned} \lim_{x\to 0} \frac{\partial g}{\partial x}(x, x) &= \lim_{x\to 0} \left[x \sin\left(\frac{1}{\sqrt{2}|x|}\right) - \frac{x^3}{2^{3/2}|x|^3} \cdot \cos\left(\frac{1}{\sqrt{2}|x|}\right) \right] \\ &= 0 - \lim_{x\to 0} \frac{x^3}{2^{3/2}|x|^3} \cdot \cos\left(\frac{1}{\sqrt{2}|x|}\right) \end{aligned}$$

The last limit is not defined.

We use the more restrictive definition of partial derivative given in what follows. This definition eliminates the possibility of obtaining a partial derivative function that is not continuous at a point in its domain.

FINAL DEFINITION 10.1: *The partial derivative with respect to x of a function $f(x, y)$ at the point (x_0, y_0) is a real number, to be denoted by the symbol $\partial f/\partial x(x_0, y_0)$, provided the limit exists:*

$$\lim_{(x,y,h)\to(x_0,y_0,0)} \frac{f(x + h, y) - f(x, y)}{h} = \frac{\partial f}{\partial x}(x_0, y_0)$$

Suppose the function $f(x, y)$ is defined on a region R. The partial derivative of $f(x, y)$ with respect to x is a function $\partial f/\partial x(x, y)$ defined on R, provided the given limit exists for any point (x_0, y_0) in R. The partial derivative with respect to y is defined analogously.

Figure 10.1 shows the δ neighborhoods for obtaining an ϵ bound for the conventional partial derivative definition, the tentative one, and the final one. With this more restrictive partial derivative definition, it is no longer possible to have a partial derivative function $\partial f/\partial x(x, y)$ defined on some region R_1, and at some boundary point (x_B, y_B) of R_1 have $\partial f/\partial x(x_B, y_B)$ exist, yet have $\lim_{(x,y)\to(x_B,y_B)} \partial f/\partial x(x, y)$ fail to exist. If $\partial f/\partial x(x_B, y_B)$ exists, this implies

$$\lim_{(x,y,h)\to(x_B,y_B,0)} \frac{f(x+h, y) - f(x, y)}{h} = \frac{\partial f}{\partial x}(x_B, y_B)$$

There is a modulus function $\delta(\epsilon)$ for this limit such that in the $(x_B, y_B, 0)$ neighborhood

$$|(x, y, h) - (x_B, y_B, 0)| < \delta(\epsilon)$$

we have

$$\left| \frac{f(x+h, y) - f(x, y)}{h} - \frac{\partial f}{\partial x}(x_B, y_B) \right| < \epsilon$$

whenever $(f(x+h, y) - f(x, y))/h$ is defined. Because (x_B, y_B) is a boundary point of R_1, there are points of R_1 in the (x_B, y_B) neighborhood $|(x, y) - (x_B, y_B)| < \delta(\epsilon)$. At any such point (x, y), we can hold x and y fixed and take the limit

$$\lim_{h\to 0} \frac{f(x+h, y) - f(x, y)}{h}$$

all the while staying in the neighborhood of $(x_B, y_B, 0)$ given in the preceding. The limit is $\partial f/\partial x(x, y)$, so we obtain the inequality $|\partial f/\partial x(x, y) - \partial f/\partial x(x_B, y_B)| \le \epsilon$. Because this inequality holds for all such points of the

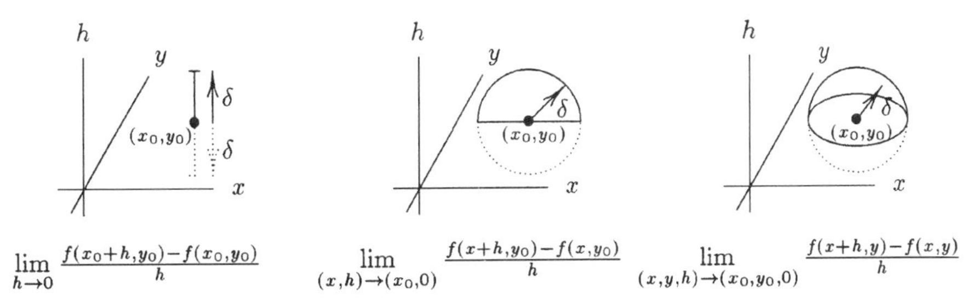

Figure 10.1. The δ neighborhoods in (x, y, h) space needed to bound three partial derivative qoutients to within ϵ of $\partial f/\partial x(x_0, y_0)$.

(x_B, y_B) neighborhood $|(x, y) - (x_B, y_B)| < \delta(\epsilon)$, this implies

$$\lim_{(x,y)\to(x_B,y_B)} \frac{\partial f}{\partial x}(x, y) = \frac{\partial f}{\partial x}(x_B, y_B)$$

Therefore, when as with $g(x, y)$ we have a partial derivative function that cannot be extended to a domain boundary point, there can be no partial derivative at the boundary point.

The partial derivative limit notation can be made more symmetric by exchanging

$$\frac{f(x+h, y) - f(x, y)}{h} \quad \text{for} \quad \frac{f(x_1, y) - f(x_2, y)}{x_1 - x_2}$$

Then the limit can be written

$$\lim_{(x_1,x_2,y)\to(x_0,x_0,y_0)} \frac{f(x_1, y) - f(x_2, y)}{x_1 - x_2} = \frac{\partial f}{\partial x}(x_0, y_0)$$

10.2 The Chain Rule

In both conventional and computable calculus, the chain rule for a function $f(g(x))$ can be used at an argument point x_0 when it is known only that $g'(x_0)$ exists and $f'(y_0)$ exists for $y_0 = g(x_0)$. In conventional calculus, with functions of two variables, in order to use at a specific argument point the various partial derivative chain rules, one needs stronger conditions than merely the existence of all derivatives at the point in question. For instance, suppose we have a function $f(u, v)$ and make the substitution $u = u(x, y)$ and $v = v(x, y)$. The functions $u(x, y)$ and $v(x, y)$ are defined in some region R, the function $f(u, v)$ is defined in some region R_1, and for all (x, y) in R, the image point $(u, v) = (u(x, y), v(x, y))$ is in R_1. Suppose we want to use at a point (x_0, x_0) in R the chain rules

$$\begin{aligned} \frac{\partial f}{\partial x} &= \frac{\partial f}{\partial u}\frac{\partial u}{\partial x} + \frac{\partial f}{\partial v}\frac{\partial v}{\partial x} \\ \frac{\partial f}{\partial y} &= \frac{\partial f}{\partial u}\frac{\partial u}{\partial y} + \frac{\partial f}{\partial v}\frac{\partial v}{\partial y} \end{aligned} \tag{10.1}$$

Besides the existence of $\partial u/\partial x$, $\partial u/\partial y$, $\partial v/\partial x$, and $\partial v/\partial y$ at (x_0, y_0), and the existence of $\partial f/\partial u$ and $\partial f/\partial v$ at the point $(u_0, v_0) = (u(x_0, y_0), v(x_0, y_0))$, in conventional calculus one needs the existence and continuity of all these derivatives in neighborhoods of the respective points given.

It is instructive to follow the conventional calculus treatment of the chain rule for this case. The functions $u(x, y)$ and $v(x, y)$ may be restricted to a neighborhood of (x_0, y_0) small enough that the image points $(u(x, y), v(x, y))$ remain in the supposed neighborhood of (u_0, v_0) where both partial derivatives $\partial f/\partial u$

and $\partial f/\partial v$ exist. We have then that

$$f(u, v) - f(u_0, v_0) = [f(u, v) - f(u_0, v)] + [f(u_0, v) - f(u_0, v_0)]$$
$$= \frac{\partial f}{\partial u}(c_u, v)(u - u_0) + \frac{\partial f}{\partial v}(u_0, c_v)(v - v_0)$$

Here we viewed the expression $f(u, v) - f(u_0, v)$ as if it were $f(u) - f(u_0)$, ignoring the unvarying v, and then used the Mean Value theorem to obtain the point c_u between u_0 and u. Similarly, we viewed the expression $f(u_0, v) - f(u_0, v_0)$ as if it were $f(v)$, ignoring the unvarying u_0, and then used the Mean Value theorem to obtain the point c_v between v_0 and v. After we make the substitutions $u = u(x, y)$ and $v = v(x, y)$, and divide by $x - x_0$, we can obtain a value for $\partial f/\partial x(u_0, v_0)$:

$$\frac{\partial f}{\partial x}(u_0, v_0) = \lim_{x \to x_0} \frac{f(u(x, y_0), v(x, y_0)) - f(u(x_0, y_0), v(x_0, y_0))}{x - x_0}$$
$$= \lim_{x \to x_0} \frac{\partial f/\partial u(c_u, v)(u - u_0) + \partial f/\partial v(u_0, c_v)(v - v_0)}{x - x_0}$$
$$= \lim_{x \to x_0} \left[\frac{\partial f}{\partial u}(c_u, v(x, y_0)) \frac{u(x, y_0) - u(x_0, y_0)}{x - x_0} + \frac{\partial f}{\partial v}(u_0, c_v) \frac{v(x, y_0) - v(x_0, y_0)}{x - x_0} \right]$$
$$= \frac{\partial f}{\partial u}(u_0, v_0)\frac{\partial u}{\partial x}(x_0, y_0) + \frac{\partial f}{\partial v}(u_0, v_0)\frac{\partial v}{\partial x}(x_0, y_0)$$

In computable calculus, with our more restrictive derivative definitions, and our relaxed limit definitions, the situation is much simpler. In order to use a particular chain rule at a point, all that is required is that all the derivatives involved in the rule exist at the point or at the appropriate image of the point. Thus in Eq. (10.1), we need only the existence of the four partial derivatives $\partial u/\partial x$, $\partial u/\partial y$ $\partial v/\partial x$ and $\partial v/\partial y$ at (x_0, y_0) and the existence of $\partial f/\partial u$ and $\partial f/\partial v$ at (u_0, v_0). To show this, we again use a remainder function as we did to prove the chain rule in Chapter 8 for functions $f(x)$. This time our remainder function involves both partial derivatives. To save space in the definition of the remainder function, and also in later equations, from now on the partial derivatives $\partial f/\partial u$, $\partial f/\partial v$ will be understood to be taken at the point (u_0, v_0), and the partial derivatives $\partial u/\partial x$, $\partial u/\partial y$, $\partial v/\partial x$, $\partial v/\partial y$ will be understood to be taken at the point (x_0, y_0). The remainder function $r(x_1, y_1, x_2, y_2)$ is defined by the equation

$$r(x_1, y_1, x_2, y_2) = f(u(x_1, y_1), v(x_1, y_1)) - f(u(x_2, y_2), v(x_2, y_2))$$
$$- \frac{\partial f}{\partial u} \cdot (u(x_1, y_1) - u(x_2, y_2)) - \frac{\partial f}{\partial v} \cdot (v(x_1, y_1) - v(x_2, y_2))$$

To prove the chain rule for $\partial f/\partial x$ we need $r(x_1, y_1, x_2, y_1)$, and to prove the chain rule for $\partial f/\partial y$ we need $r(x_1, y_1, x_1, y_2)$. The two cases are similar, and we give the proof only for $\partial f/\partial x$:

$$\left.\frac{\partial f(u(x,y), v(x,y))}{\partial x}\right|_{(x_0,y_0)}$$

$$= \lim_{(x_1,x_2,y_1)\to(x_0,x_0,y_0)} \frac{f(u(x_1,y_1), v(x_1,y_1)) - f(u(x_2,y_1), v(x_2,y_1))}{x_1 - x_2}$$

$$= \lim_{(x_1,x_2,y_1)\to(x_0,x_0,y_0)} \frac{\partial f/\partial u(u(x_1,y_1) - u(x_2,y_1)) + \partial f/\partial v(v(x_1,y_1) - v(x_2,y_1)) + r(x_1,y_1,x_2,y_1)}{x_1 - x_2}$$

$$= \frac{\partial f}{\partial u}\frac{\partial u}{\partial x} + \frac{\partial f}{\partial v}\frac{\partial v}{\partial x} + \lim_{(x_1,x_2,y_1)\to(x_0,x_0,y_0)} \frac{r(x_1,y_1,x_2,y_1)}{x_1 - x_2}$$

The limit remaining involves the function

$$\frac{r(x_1,y_1,x_2,y_1)}{x_1 - x_2}$$

$$= \frac{f(u(x_1,y_1), v(x_1,y_1)) - f(u(x_2,y_1), v(x_2,y_1)) - \partial f/\partial u(u(x_1,y_1) - u(x_2,y_1)) - \partial f/\partial v(v(x_1,y_1) - v(x_2,y_1))}{x_1 - x_2}$$

There are various cases to consider. If $u(x_1, y_1) \neq u(x_2, y_1)$ and $v(x_1, y_1) \neq v(x_2, y_1)$, then the expression part $(f(u(x_1, y_1), v(x_1, y_1)) - f(u(x_2, y_1), v(x_2, y_1)))/(x_1 - x_2)$ can be written as

$$\frac{f(u(x_1,y_1), v(x_1,y_1)) - f(u(x_2,y_1), v(x_1,y_1)) + f(u(x_2,y_1), v(x_1,y_1)) - f(u(x_2,y_1), v(x_2,y_1))}{x_1 - x_2}$$

$$= \frac{f(u(x_1,y_1), v(x_1,y_1)) - f(u(x_2,y_1), v(x_1,y_1))}{u(x_1,y_1) - u(x_2,y_1)} \cdot \frac{u(x_1,y_1) - u(x_2,y_1)}{x_1 - x_2}$$

$$+ \frac{f(u(x_2,y_1), v(x_1,y_1)) - f(u(x_2,y_1), v(x_2,y_1))}{v(x_1,y_1) - v(x_2,y_1)} \cdot \frac{v(x_1,y_1) - v(x_2,y_1)}{x_1 - x_2}$$

The limit of this is $(\partial f/\partial u)(\partial u/\partial x) + (\partial f/\partial v)(\partial v/\partial x)$, while the limit of the remaining part of the expression is $-(\partial f/\partial u)(\partial u/\partial x) - (\partial f/\partial v)(\partial v/\partial x)$, giving a limit for the whole expression of zero. If $u(x_1, y_1) = u(x_2, y_1)$ and $v(x_1, y_1) = v(x_2, y_1)$, then $r(x_1, y_1, x_2, y_1)$ is zero and the limit is zero. In an intermediate case, say $u(x_1, y_1) \neq u(x_2, y_1)$ and $v(x_1, y_1) = v(x_2, y_1)$, the entire expression is

$$\frac{f(u(x_1,y_1), v(x_1,y_1)) - f(u(x_2,y_1), v(x_1,y_1)) - \partial f/\partial u(u(x_1,y_1) - u(x_2,y_1))}{x_1 - x_2}$$

$$= \frac{f(u(x_1,y_1), v(x_1,y_1)) - f(u(x_2,y_1), v(x_1,y_1))}{u(x_1,y_1) - u(x_2,y_1)} \cdot \frac{u(x_1,y_1) - u(x_2,y_1)}{x_1 - x_2}$$

$$- \frac{\partial f}{\partial u} \cdot \frac{u(x_1,y_1) - u(x_2,y_1)}{x_1 - x_2}$$

giving a limit of $(\partial f/\partial u)(\partial u/\partial x) - (\partial f/\partial u)(\partial u/\partial x) = 0$. Thus regardless of which case occurs, the limit is zero. A limit modulus constructed from the given moduli, but assuming the first case, would apply to all cases. We list our chain rule result as:

THEOREM 10.1: Suppose $u(x, y)$ and $v(x, y)$ are functions defined on a region R with the image points $(u(x, y), v(x, y))$ contained in a region R_1. If both functions have both partial derivatives at a point (x_0, y_0) in R, the image point being (u_0, v_0) in R_1, and $f(u, v)$ is a function defined on R_1 having both partial derivatives at (u_0, v_0), then the function $g(x, y) = f(u(x, y), v(x, y))$ has both partial derivatives at (x_0, y_0) and these are

$$\frac{\partial g}{\partial x}(x_0, y_0) = \frac{\partial f}{\partial u}(u_0, v_0)\frac{\partial u}{\partial x}(x_0, y_0) + \frac{\partial f}{\partial v}(u_0, v_0)\frac{\partial v}{\partial x}(x_0, y_0)$$
$$\frac{\partial g}{\partial y}(x_0, y_0) = \frac{\partial f}{\partial u}(u_0, v_0)\frac{\partial u}{\partial y}(x_0, y_0) + \frac{\partial f}{\partial v}(u_0, v_0)\frac{\partial v}{\partial y}(x_0, y_0)$$

There is no difficulty extending the chain rule to other cases similarly, and requiring only that all chain rule partial derivatives (or ordinary derivatives) exist at a single point. For instance, the directional deriviative of a function $f(x, y)$ is frequently used:

DEFINITION 10.2: *The directional derivative of $f(x, y)$ at a point (x_0, y_0) in the direction $\theta = \theta_0$,*

$$\textit{written} \quad D_{\theta_0} f(x_0, y_0)$$

is the partial derivative with respect to r of the function

$$g(x, y, r, \theta) = f(x + r\cos\theta, y + r\sin\theta)$$

at the point $(x_0, y_0, 0, \theta_0)$.

We obtain by an appropriate chain rule the equation

$$D_{\theta_0} f(x_0, y_0) = \frac{\partial f}{\partial x}(x_0, y_0)\cos\theta_0 + \frac{\partial f}{\partial y}(x_0, y_0)\sin\theta_0$$

The only condition needed for a directional derivative to exist at (x_0, y_0) is that the two partial derivatives $\partial f/\partial x(x_0, y_0)$ and $\partial f/\partial y(x_0, y_0)$ exist.

10.3 Equality of Cross Derivatives

In conventional calculus, because a derivative can be assigned a discontinuous value, it is not always true that $\partial^2 f/\partial x\partial y$ equals $\partial^2 f/\partial y\partial x$. Here is an example

([38], page 42) where there is inequality:

$$f(x, y) = \begin{cases} 2xy\frac{x^2-y^2}{x^2+y^2} & \text{if } (x, y) \neq (0, 0) \\ 0 & \text{if } (x, y) = (0, 0) \end{cases}$$

(By Theorem 6.11 the function $f(x, y)$ is a computable function defined everywhere in the xy plane.) According to the conventional calculus definition of partial derivatives, the first derivatives are

$$\frac{\partial f}{\partial x}(x, y) = \begin{cases} 2y\frac{x^2-y^2}{x^2+y^2} + 2xy\frac{4xy^2}{(x^2+y^2)^2} & \text{if } (x, y) \neq (0, 0) \\ 0 & \text{if } (x, y) = (0, 0) \end{cases}$$

$$\frac{\partial f}{\partial y}(x, y) = \begin{cases} 2x\frac{x^2-y^2}{x^2+y^2} - 2xy\frac{4x^2y}{(x^2+y^2)^2} & \text{if } (x, y) \neq (0, 0) \\ 0 & \text{if } (x, y) = (0, 0) \end{cases}$$

For the second derivatives, we find $\partial^2 f/\partial x\partial y(0, 0) \neq \partial^2 f/\partial y\partial x(0, 0)$.

$$\frac{\partial^2 f}{\partial x\partial y}(0, 0) = \lim_{h\to 0} \frac{\frac{\partial f}{\partial y}(0+h, 0) - \frac{\partial f}{\partial y}(0, 0)}{h} = \lim_{h\to 0} \frac{2h - 0}{h} = 2$$

$$\frac{\partial^2 f}{\partial y\partial x}(0, 0) = \lim_{h\to 0} \frac{\frac{\partial f}{\partial x}(0, 0+h) - \frac{\partial f}{\partial y}(0, 0)}{h} = \lim_{h\to 0} \frac{-2h - 0}{h} = -2$$

In conventional calculus, in order that the partial derivative $\partial^2 f/\partial x\partial y$ be equal to the partial derivative $\partial^2 f/\partial y\partial x$ at a point (x_0, y_0), one needs both derivatives to exist and be continuous in a neighborhood of (x_0, y_0).

Let us examine a conventional calculus proof of this, the proof also valid in computable calculus. We assume that in the neighborhood N_0 of (x_0, y_0) the functions $f(x, y)$, $\partial f/\partial x(x, y)$, $\partial f/\partial y(x, y)$, $\partial^2 f/\partial x\partial y(x, y)$, and $\partial^2 f/\partial y\partial x(x, y)$ are all defined. In computable calculus these functions are necessarily continuous in N_0, but in conventional calculus this must be added as an additional assumption.

The proof consists in evaluating a limit in two different ways:

$$\lim_{(x,y)\to(x_0,y_0)} \frac{f(x, y) - f(x, y_0) - f(x_0, y) + f(x_0, y_0)}{(x - x_0)(y - y_0)} \tag{10.2}$$

First note that in conventional calculus, this limit cannot exist, because the given quotient is not defined on the (x, y) points of the lines $x = x_0$ and $y = y_0$ going through (x_0, y_0). This can be rectified by defining the quotient on the two lines to equal the desired limit value. This difficulty does not occur with the relaxed limit definition of computable calculus.

Assume (x, y) is in N_0. First write the displayed numerator as $g(x) - g(x_0)$ where $g(x) = f(x, y) - f(x, y_0)$, and apply the Mean Value theorem to replace it by

$$g'(c_x)(x - x_0) = \left[\frac{\partial f}{\partial x}(c_x, y) - \frac{\partial f}{\partial x}(c_x, y_0)\right](x - x_0)$$

Here c_x is a point between x and x_0. Then considering $\partial f/\partial x(c_x, y) - \partial f/\partial x(c_x, y_0)$ to be a function of y, apply the Mean Value theorem a second time to replace this expression by $\partial^2 f/\partial y \partial x(c_x, c_y)(y - y_0)$. This time c_y is a point between y and y_0. This converts Eq. (10.2) to

$$\lim_{(x,y)\to(x_0,y_0)} \frac{\partial^2 f}{\partial y \partial x}(c_x, c_y) = \frac{\partial^2 f}{\partial y \partial x}(x_0, y_0)$$

Next write the numerator in the form $h(y) - h(y_0)$ where $h(y) = f(x, y) - f(x_0, y)$ and apply the Mean Value theorem to replace it by

$$h'(\widehat{c}_y)(y - y_0) = \left[\frac{\partial f}{\partial y}(x, \widehat{c}_y) - \frac{\partial f}{\partial y}(x_0, \widehat{c}_y)\right](y - y_0)$$

Then consider $\partial f/\partial y(x, \widehat{c}_y) - \partial f/\partial y(x_0, \widehat{c}_y)$ to be a function of x and apply the Mean Value theorem a second time to replace this expression by $\partial^2 f/\partial x \partial y(\widehat{c}_x, \widehat{c}_y)(x - x_0)$. This converts Eq. (10.2) to

$$\lim_{(x,y)\to(x_0,y_0)} \frac{\partial^2 f}{\partial x \partial y}(\widehat{c}_x, \widehat{c}_y) = \frac{\partial^2 f}{\partial x \partial y}(x_0, y_0)$$

Equating the two limits we obtain $\partial^2 f/\partial y \partial x(x_0, y_0) = \partial^2 f/\partial x \partial y(x_0, y_0)$.

In computable calculus, we can prove this result under simpler conditions:

THEOREM 10.2: Suppose $f(x, y)$ is defined and has both first partial derivatives in a neighborhood N of the point (x_0, y_0). If one of the two second derivatives $\partial^2 f/\partial x \partial y(x_0, y_0)$, $\partial^2 f/\partial y \partial x(x_0, y_0)$ exists, then the other exists also and has the same value.

We assume $\partial^2 f/\partial y \partial x(x_0, y_0)$ exists, so there is a limit modulus $\delta(\epsilon)$ such that

$$|(x_1, y_1, y_2) - (x_0, y_0, y_0)| < \delta(\epsilon) \tag{10.3}$$

implies for the case $y_1 \neq y_2$ that

$$\left| \frac{\partial f/\partial x(x_1, y_1) - \partial f/\partial x(x_1, y_2)}{y_1 - y_2} - \frac{\partial^2 f}{\partial y \partial x}(x_0, y_0) \right| < \epsilon \tag{10.4}$$

The neighborhood N is defined by some inequality $|(x, y) - (x_0, y_0)| < \delta_0$, and we may assume that the $\delta(\epsilon)$ value is always small enough so that if we have inequality (10.3), then the points (x_1, y_1) and (x_1, y_2) are in N. If need be, we can replace $\delta(\epsilon)$ by $\min(\delta_0, \delta(\epsilon))$ to ensure that this occurs. Here note that

$$\begin{aligned}|(x_1, y_1) - (x_0, y_0)| &= \sqrt{(x_1 - x_0)^2 + (y_1 - y_0)^2} \\ &\leq \sqrt{(x_1 - x_0)^2 + (y_1 - y_0)^2 + (y_2 - y_0)^2} \\ &= |(x_1, y_1, y_2) - (x_0, y_0, y_0)|\end{aligned}$$

This implies that if inequality (10.3) holds, then (x_1, y_1) is in N. Similarly, the point (x_1, y_2) also is in N. For use later on, note that $|(x_1, y_1) - (x_0, y_0)| < \delta$ and $|z_1 - z_0| < \delta$ implies $|(x_1, y_1, z_1) - (x_0, y_0, z_0)| < \sqrt{2}\delta$, because

$$\begin{aligned}|(x_1, y_1, z_1) - (x_0, y_0, z_0)| &= \sqrt{[(x_1 - x_0)^2 + (y_1 - y_0)^2] + (z_1 - z_0)^2} \\ &< \sqrt{\delta^2 + \delta^2} = \sqrt{2}\delta\end{aligned}$$

The general plan is to show that for the limit modulus $\delta_1(\epsilon)$ equal to $\frac{1}{2}\delta(\epsilon/2)$, the inequality

$$|(x_1, x_2, y_1) - (x_0, x_0, y_0)| < \delta_1(\epsilon)$$

implies for the case $x_1 \neq x_2$ that

$$\left| \frac{\partial f/\partial y(x_1, y_1) - \partial f/\partial y(x_2, y_1)}{x_1 - x_2} - \frac{\partial^2 f}{\partial y \partial x}(x_0, y_0) \right| < \epsilon \tag{10.5}$$

This shows that the derivative $\partial^2 f/\partial x \partial y(x_0, y_0)$ exists and equals $\partial^2 f/\partial y \partial x(x_0, y_0)$.

The first step is to substitute for the terms $\partial f/\partial y(x_1, y_1)$ and $\partial f/\partial y(x_2, y_1)$, which appear on the left-hand side of the inequality (10.5), the quotients

$$\frac{f(x_1, y_1) - f(x_1, y_3)}{y_1 - y_3} \quad \text{and} \quad \frac{f(x_2, y_1) - f(x_2, y_3)}{y_1 - y_3}$$

respectively, where the coordinate y_3 satisfies the inequality $|y_3 - y_0| < \delta_1(\epsilon)$. We choose y_3 so near y_1 that the difference between each original quantity and its replacement is $< (\epsilon/4)|x_1 - x_2|$. These substitutions change the left-hand side quantity in Eq. (10.5) to the quantity

$$\left| \frac{f(x_1, y_1) - f(x_1, y_3) - f(x_2, y_1) + f(x_2, y_3)}{(x_1 - x_2)(y_1 - y_3)} - \frac{\partial^2 f}{\partial x \partial y}(x_0, y_0) \right| \tag{10.6}$$

and the error made by this change is less than $\epsilon/2$.

Next we note that the function $g(x) = f(x, y_1) - f(x, y_3)$ is such that $g(x_1) - g(x_2)$ equals the numerator of the quotient in Eq. (10.6). The derivative of $g(x)$ is $\partial f/\partial x(x, y_1) - \partial f/\partial x(x, y_3)$. Because we have $|(x_1, x_2, y_1) - (x_0, x_0, y_0)| < \delta_1(\epsilon)$, which implies both inequalities $|(x_1, y_1) - (x_0, y_0)| < \delta_1(\epsilon)$ and $|(x_2, y_1) - (x_0, y_0)| < \delta_1(\epsilon)$, and because $|y_3 - y_0| < \delta_1(\epsilon)$, it is certain that both quantities $|(x_1, y_1, y_3) - (x_0, y_0, y_0)|$ and $|(x_2, y_1, y_3) - (x_0, y_0, y_0)|$ are $< \sqrt{2} \cdot \delta_1(\epsilon) < 2\delta_1(\epsilon) = \delta(\epsilon/2)$. Therefore, for all x between x_1 and x_2, we can use the relationship of Eq. (10.4) to bound the $g(x)$ derivative by the inequalities

$$-\frac{\epsilon}{2} < \frac{\partial f/\partial x(x, y_1) - \partial f/\partial x(x, y_3)}{y_1 - y_3} - \frac{\partial^2 f}{\partial y \partial x}(x_0, y_0) < \frac{\epsilon}{2} \tag{10.7}$$

Now we apply the Mean Value theorem to $g(x)$ over the interval from x_1 to x_2, and obtain

$$\frac{g(x_1) - g(x_2)}{x_1 - x_2} = \frac{\partial f}{\partial x}(c_x, y_1) - \frac{\partial f}{\partial x}(c_x, y_3)$$

After we divide this equation by $y_1 - y_3$ and subtract $\partial^2 f/\partial y \partial x(x_0, y_0)$, we can use Eq. (10.7) to obtain the inequalities

$$-\frac{\epsilon}{2} < \frac{g(x_1) - g(x_2)}{(x_1 - x_2)(y_1 - y_3)} - \frac{\partial^2 f}{\partial y \partial x}(x_0, y_0) < \frac{\epsilon}{2}$$

After we substitute for $g(x_1)$ and $g(x_2)$, we obtain the inequalities

$$-\frac{\epsilon}{2} < \frac{f(x_1, y_1) - f(x_1, y_3) - f(x_2, y_1) + f(x_2, y_3)}{(x_1 - x_2)(y_1 - y_2)} - \frac{\partial^2 f}{\partial y \partial x}(x_0, y_0) < \frac{\epsilon}{2}$$

This shows that the quantity (10.6) is $< \epsilon/2$, and taking into account the previous $\epsilon/2$ discrepancy gives us inequality (10.5).

CHAPTER 11

The Differential Equation $y' = f(x, y)$

11.1 Introduction

In this chapter we consider the problem of solving the differential equation $y' = f(x, y)$, where $f(x, y)$ is uniformly continuous in some bounded region. The specific problem we consider, usually called the initial value problem for a first order differential equation, is the following: We are given the value y_0 of the desired solution $y(x)$ at some x value a, and we want the solution $y(x)$ to the differential equation in an interval $[a, b]$. That is, in the interval $[a, b]$ the function $y(x)$ must satisfy the equation

$$y'(x) = f(x, y(x))$$

and also satisfy the initial value condition $y(a) = y_0$. We presume the function $f(x, y)$ to be uniformly continuous in any rectangle R that is defined by the inequalities

$$a \leq x \leq b \qquad |y - y_0| \leq M$$

There are additional conditions that must be imposed in order to ensure that a unique solution to this problem exists. For example, certain functions $f(x, y)$ allow the solution to grow or to decrease too rapidly, so that at some point c in $[a, b]$ we find $\lim_{x \to c^-} y(x)$ is ∞ or $-\infty$, so a solution throughout $[a, b]$

is not possible. For instance, the differential equation $y' = y^2$, with the initial value condition $y(0) = 1$, has the solution $y(x) = 1/(1 - x)$, so $\lim_{y \to 1^-} y(x) = \infty$, and a solution in the interval $[0, 2]$ is not possible.

The possibility of this difficulty can be eliminated by finding positive quantities M_1 and M_2 such that for x in the interval $[a, b]$ we have $|f(x, y)| < M_1 + M_2|y|$. Notice that for the example just given, $f(x, y)$ equals y^2 and cannot be bounded this way. When this type of bound for $f(x, y)$ is obtained, a solution function $y(x)$ in $[a, b]$ cannot approach ∞ or $-\infty$, as we now show. Define the two functions $y_1(x)$ and $y_2(x)$ by the equations

$$\begin{aligned} y_1'(x) &= M_1 + M_2 y, & y_1(a) &= |y_0| + 1 \\ y_2'(x) &= -M_1 - M_2 y, & y_2(a) &= -|y_0| - 1 \end{aligned}$$

These functions are defined in $[a, b]$ and can be found, if required, by the usual method of solving a differential equation by separation of variables. For instance, the solution in $[1, 10]$ to the equation $y'(x) = 3 + 4y$, with the initial condition $y(1) = 5$, can be obtained this way:

$$\begin{aligned} \int_5^y \frac{dy}{3 + 4y} &= \int_1^x dx \\ \frac{1}{4} \ln \frac{3 + 4y}{23} &= x - 1 \\ \frac{3 + 4y}{23} &= e^{4(x-1)} \\ y &= \frac{1}{4}(23e^{4(x-1)} - 3) \end{aligned}$$

A solution $y(x)$ to the initial value problem $y'(x) = f(x, y)$, $y(a) = y_0$, must satisfy the bounds $y_2(x) < y(x) < y_1(x)$ for all x in $[a, b]$. This is because $y_2(a) < y(a) < y_1(a)$, so for x near a, the given bounds hold. And if the curve $y = y(x)$ crossed or touched either of the curves $y = y_1(x)$ or $y = y_2(x)$ at some argument x' to the right of a, we violate the derivative relation $y_2'(x) < y'(x) < y_1'(x)$ at x'. From the two bounding functions $y_1(x)$ and $y_2(x)$, we can determine a positive bound M such that for x in $[a, b]$ a solution $y(x)$ must satisfy $-M < y(x) < M$. The bound M can simply be taken as $\max(|y_1(b)|, |y_2(b)|)$. We call the rectangle $R_{[a,b],M}$ defined by the inequalities

$$a \le x \le b \qquad -M \le y \le M$$

a *containment rectangle* for the initial value problem $y' = f(x, y)$, $y(a) = y_0$. For this initial value problem, we need only consider the behavior of $f(x, y)$ within the containment rectangle.

A second possible difficulty is illustrated by the initial value problem

$$y' = y^{1/3}, \qquad y(0) = y_0$$

If y_0 is unequal to zero, there is a single solution, but if y_0 is zero, there is an infinite number of solutions in the interval [0, 1]. When y_0 is zero, we may take $y(x)$ equal to 0 everywhere, as this clearly is a solution, or we may solve the differential equation in the usual way by separating variables:

$$\int_{y_0}^{y} \frac{dy}{y^{1/3}} = \int_0^x dx$$

$$\tfrac{3}{2}\left[y^{2/3} - y_0^{2/3}\right] = x$$

$$y = \left[\pm\sqrt{y_0^{2/3} + \tfrac{2}{3}x}\,\right]^3$$

If y_0 is unequal to zero, only one sign of the two shown in the preceding is valid, because at $x=0$ we must obtain $y(0)=y_0$. (This justifies our assertion that if y_0 is unequal to 0, there is only one solution.) However, if y_0 is zero, both signs are possible, so we obtain two additional solutions besides the zero solution, namely $[\frac{2}{3}x]^{3/2}$ and $-[\frac{2}{3}x]^{3/2}$. Besides these three solutions, there are an infinite number of other solutions. The other solutions are obtained by requiring $y(x)$ to equal the first solution, that is, 0, up to some point $x=c$ in (0, 1), whereupon $y(x)$ switches to either the second or the third solution. Specifically, choosing any c in the interval (0, 1), and choosing a plus or a minus sign, we have the solution

$$y(x) = \begin{cases} 0 & \text{if } 0 \le x \le c \\ (\text{sign})[\frac{2}{3}(x-c)]^{3/2} & \text{if } c < x \le 1 \end{cases}$$

Notice that there is no discontinuity of either $y(x)$ or $y'(x)$ at $x=c$.

In the next section we show that if we require $f(x, y)$ to be uniformly continuous within a containment rectangle and satisfy there a certain condition called the Lipschitz condition, then there is a unique solution to the initial value problem. In the final section of this chapter, we show that if we require only that $f(x, y)$ be uniformly continuous in a containment rectangle, it is possible for the initial value problem to have no solution.

11.2 The Lipschitz Condition

The function $f(x, y)$ satisfies the Lipschitz condition in a containment rectangle $R_{[a,b],M}$ if there is a positive constant L such that for any two containment rectangle points (x, y_1) and (x, y_2) with the same x coordinate, we have

$$|f(x, y_1) - f(x, y_2)| < L|y_1 - y_2|$$

When we require that $f(x, y)$ satisfy the Lipschitz condition in a containment rectangle and be uniformly continuous in the rectangle, multiple solutions to the initial value problem no longer are possible.

If the partial derivative $\partial f/\partial y(x, y)$ exists in the containment rectangle, and we can find a positive constant L such that the inequality $|\partial f/\partial y(x, y)| < L$ holds in the rectangle, then the Lipschitz condition is satisfied there. We obtain the Lipschitz condition by applying the Mean Value theorem to the function $g(y) = f(x, y)$, where here x is held fixed. We have then

$$\begin{aligned}|f(x, y_1) - f(x, y_2)| &= |g(y_1) - g(y_2)| = |g'(c_y)(y_1 - y_2)| \\ &= \left|\frac{\partial f}{\partial y}(x, c_y)\right||y_1 - y_2| < L|y_1 - y_2|\end{aligned}$$

Note that the function $f(x, y)$ of the preceding example is $y^{1/3}$, and at $y = 0$ this function does not have a partial derivative with respect to y.

THEOREM 11.1: The initial value problem $y' = f(x, y)$, $y(a) = y_0$ has a unique, uniformly continuous solution $y(x)$ in $[a, b]$ if the function $f(x, y)$ is uniformly continuous in a containment rectangle $R_{[a,b],M}$ and satisfies there the Lipschitz condition.

We prove the result in two steps. First, because $f(x, y)$ is uniformly continuous, we can obtain a sequence of uniformly continuous functions $y_n(x)$ satisfying for x in $[a, b]$ the conditions

$$|y_n'(x) - f(x, y_n(x))| < 10^{-n}, \qquad y_n(a) = y_0 \tag{11.1}$$

Then because of the Lipschitz condition, we show that the sequence of functions $y_n(x)$ is uniformly Cauchy on $[a, b]$ and converges there to a uniformly continuous limit function $y(x)$ that satisfies the initial value problem.

The construction of $y_n(x)$ is done as follows: Because $f(x, y)$ is uniformly continuous in $R_{[a,b],M}$, there is a modulus $\delta(\epsilon)$ such that if (x_1, y_1) and (x_2, y_2) are any two points in $R_{[a,b],M}$,

$$|(x_1, y_1) - (x_2, y_2)| < \delta(\epsilon) \quad \text{implies} \quad |f(x_1, y_1) - f(x_2, y_2)| < \epsilon$$

By generalizing the method used in the proof of Theorem 7.1, we can find a positive integer $N > |f(x, y)|$ in $R_{[a,b],M}$ with the aid of the function $\delta(\epsilon)$. We find another positive integer N_n such that $\sqrt{1 + (5N)^2} \cdot (b - a)/N_n < \delta(10^{-n})$. Denote the quantity $(b - a)/N_n$ by the symbol $\Delta^{(n)}$. We determine in $R_{[a,b],M}$ the finite sequence of points $(x_i^{(n)}, y_i^{(n)})$ for $0 \le i \le N_n$ as follows:

$$(x_i^{(n)}, y_i^{(n)}) = \begin{cases} (a, y_0) & \text{if } i = 0 \\ (x_{i-1}^{(n)}, y_{i-1}^{(n)}) + (\Delta^{(n)}, f(x_{i-1}^{(n)}, y_{i-1}^{(n)})\Delta^{(n)}) & \text{if } 0 < i \le N_n \end{cases}$$

Notice that for any two successive points $(x_i^{(n)}, y_i^{(n)})$ and $(x_{i+1}^{(n)}, y_{i+1}^{(n)})$, we have

$$\left|(x_i^{(n)}, y_i^{(n)}) - (x_{i+1}^{(n)}, y_{i+1}^{(n)})\right| = \left|(\Delta^{(n)}, f(x_i^{(n)}, y_i^{(n)})\Delta^{(n)})\right| < \sqrt{1 + N^2}\,\Delta^{(n)} < \delta(10^{-n})$$

Therefore $f(x, y)$ varies by $< 10^{-n}$ when evaluated along a straight line segment joining these two successive points. If we use all these line segments joining successive points, making a broken line path from the first point (a, y_0) all the way to the last point $(b, y_{N_n}^{(n)})$, and define $y_n(x)$ to be the function which for any x equals the y value obtained by intersecting the constant x line with the path, we would have a function satisfying our requirements, except for one deficiency. The derivative $y_n'(x_i^{(n)})$ is not defined if two line segments with differing slopes meet at the point $(x_i^{(n)}, y_i^{(n)})$. We correct this flaw by allowing the path between two successive points to be curved: In the x interval $[x_i^{(n)}, x_{i+1}^{(n)}]$ of width $\Delta^{(n)}$, the function $y_n(x)$ previously was defined as

$$y_i^{(n)} + m_i(x - x_i^{(n)})$$

where m_i equals $f(x_i^{(n)}, y_i^{(n)})$. Now $y_n(x)$ is defined to be

$$y_i^{(n)} + m_i(x - x_i^{(n)}) + (m_i - m_{i+1})\left(\frac{(x - x_i^{(n)})^2}{\Delta^{(n)}} - \frac{(x - x_i^{(n)})^3}{(\Delta^{(n)})^2}\right)$$

The slope along this path, instead of being fixed at m_i as occurred previously, is now

$$m_i + (m_i - m_{i+1})\left(\frac{2(x - x_i^{(n)})}{\Delta^{(n)}} - \frac{3(x - x_i^{(n)})^2}{(\Delta^{(n)})^2}\right)$$

This slope equals m_i at $x = x_i^{(n)}$, and equals m_{i+1} at $x = x_{i+1}^{(n)}$, so no longer can there be two slopes at the points $(x_i^{(n)}, y_i^{(n)})$. With this system, the successive points $(x_i^{(n)}, y_i^{(n)})$ define a broken line path, but the function $y_n(x)$ no longer uses this path, although it still goes through the successive points. In the x interval $[x_i^{(n)}, x_{i+1}^{(n)}]$, the added $y_n(x)$ term has magnitude no greater than $(|m_i| + |m_{i+1}|)(\Delta^{(n)} + \Delta^{(n)}) < 4N\Delta^{(n)}$, and this implies that if (x, y) is any point on the curved path, we have

$$|(x_i^{(n)}, y_i^{(n)}) - (x, y)| < \sqrt{1 + (5N)^2}\,\Delta^{(n)} < \delta(10^{-n})$$

so $f(x, y)$ still varies by $< 10^{-n}$ when evaluated along the curved path.

Next we show that the functions $y_n(x)$ converge to a limit. Using Eq. (11.1), we have

$$\begin{aligned}
&|y_n'(x) - y_m'(x)| \\
&\quad = |y_n'(x) - f(x, y_n(x)) + f(x, y_n(x)) - f(x, y_m(x)) + f(x, y_m(x)) - y_m'(x)| \\
&\quad \le |y_n'(x) - f(x, y_n(x))| + |f(x, y_n(x)) - f(x, y_m(x))| + |f(x, y_m(x)) - y_m'(x)| \\
&\quad \le 10^{-n} + |f(x, y_n(x)) - f(x, y_m(x))| + 10^{-m}
\end{aligned}$$

The Lipschitz condition can now be used:

$$|y_n'(x) - y_m'(x)| \leq L|y_n(x) - y_m(x)| + 10^{-n} + 10^{-m} \tag{11.2}$$

We also have

$$y_n(x) - y_m(x) = y_n(a) - y_m(a) + \int_a^x (y_n'(t) - y_m'(t))\, dt$$

Because $y_n(a) = y_m(a) = y_0$, after taking absolute values, we have

$$|y_n(x) - y_m(x)| \leq \int_a^x |y_n'(t) - y_m'(t)|\, dt$$

Using Eq. (11.2), this gives

$$|y_n(x) - y_m(x)| \leq \int_a^x L|y_n(t) - y_m(t)|\, dt + (10^{-n} + 10^{-m})(x - a) \tag{11.3}$$

Now set $h(x)$ equal to $\int_a^x |y_n(t) - y_m(t)|\, dt$. The inequality given in the preceding may be written as

$$h'(x) - Lh(x) \leq (10^{-n} + 10^{-m})(x - a)$$

Multiply this inequality by the positive quantity $e^{-L(x-a)}$ to get

$$(h(x)e^{-L(x-a)})' \leq (10^{-n} + 10^{-m})(x - a)e^{-L(x-a)}$$

We have then

$$\begin{aligned} h(x)e^{-L(x-a)} &= \int_a^x \frac{d}{dt}(h(t)e^{-L(t-a)})dt \\ &\leq \int_a^x (10^{-n} + 10^{-m})(t - a)e^{-L(t-a)}dt \\ &= (10^{-n} + 10^{-m}) \left[(t - a)\frac{e^{-L(t-a)}}{-L} - \frac{e^{-L(t-a)}}{L^2} \right]_a^x \\ &= (10^{-n} + 10^{-m}) \left[(x - a)\frac{e^{-L(x-a)}}{-L} + \frac{1 - e^{-L(x-a)}}{L^2} \right] \end{aligned}$$

When we multiply by $e^{L(x-a)}$, we obtain

$$h(x) \leq \frac{(10^{-n} + 10^{-m})}{L} \left[\frac{e^{L(x-a)} - 1}{L} - (x - a) \right]$$

When we substitute in inequality (11.3), we finally get

$$|y_n(x) - y_m(x)| \leq (10^{-n} + 10^{-m}) \cdot \frac{e^{L(x-a)} - 1}{L} \tag{11.4}$$

The last inequality implies that $y_n(x)$ is a uniformly Cauchy sequence of functions, as we now show. The maximum of the nonnegative function

$e^{L(x-a)} - 1$ in the interval $[a, b]$ is $e^{L(b-a)} - 1$. If we let M denote the number $(e^{L(b-a)} - 1)/L$, we can take $\log_{10}(2M/\epsilon)$ as $N_C(\epsilon)$, because $n > \log_{10}(2M/\epsilon)$ implies $10^n > (2M/\epsilon)$ or $10^{-n} < (\epsilon/2M)$, so if $n, m > \log_{10}(2M/\epsilon) = N_C(\epsilon)$, we obtain $|y_n(x) - y_m(x)| < \epsilon$, the requirement for a uniform Cauchy sequence of functions. Theorem 6.10 now implies that $y_n(x)$ converges uniformly on $[a, b]$ to a limit function $y(x)$.

The functions $y_n(x)$ are also uniformly continuous on $[a, b]$, because each function is a piecewise polynomial function on $[a, b]$. Theorem 7.2 now implies that the limit function $y(x)$ is uniformly continuous on $[a, b]$. Now consider the function $z(x)$ defined by the equation

$$z(x) = y_0 + \int_a^x f(t, y(t))\, dt$$

We show that for any x in $[a, b]$ and for any positive number ϵ, we have $|z(x) - y(x)| < M_0\epsilon$, where M_0 is a positive constant, implying $z(x) = y(x)$ on $[a, b]$. Then according to Theorem 9.3, the equation $y(x) = y_0 + \int_a^x f(t, y(t))\, dt$ can be differentiated to yield $y'(x) = f(x, y(x))$, so $y(x)$ is the solution to the initial value problem. To prove the specified inequality, choose an n such that $|y_n(x) - y(x)| < \epsilon$ and such that $|y_n'(x) - f(x, y_n(x))| < \epsilon$. Then

$$\begin{aligned}
|z(x) - y(x)| &= |z(x) - y_n(x) + y_n(x) - y(x)| \\
&\le |z(x) - y_n(x)| + |y_n(x) - y(x)| \\
&= \left| y_0 + \int_a^x f(t, y(t))\, dt - y_0 - \int_a^x y_n'(t)\, dt \right| + |y_n(x) - y(x)| \\
&= \left| \int_a^x \big(f(t, y(t)) - y_n'(t)\big)\, dt \right| + |y_n(x) - y(x)| \\
&= \left| \int_a^x \big(f(t, y(t)) - f(t, y_n(t)) + f(t, y_n(t)) - y_n'(t)\big)\, dt \right| \\
&\quad + |y_n(x) - y(x)| \\
&\le \int_a^x |f(t, y(t)) - f(t, y_n(t)) + f(t, y_n(t)) - y_n'(t)|\, dt \\
&\quad + |y_n(x) - y(x)| \\
&\le \int_a^x |f(t, y(t)) - f(t, y_n(t))| dt + \int_a^x |f(t, y_n(t)) - y_n'(t)| dt \\
&\quad + |y_n(x) - y(x)| \\
&< \int_a^x L|y(t)) - y_n(t)| dt + \int_a^x \epsilon\, dt + |y_n(x) - y(x)| \\
&< L\epsilon(x - a) + \epsilon(x - a) + \epsilon \le (L + 1)\epsilon(b - a) + \epsilon \\
&= [(L + 1)(b - a) + 1]\epsilon
\end{aligned}$$

The constant M_0 is $(L + 1)(b - a) + 1$.

To show that there cannot be two distinct solutions to the initial value problem, assume a second solution $\widehat{y}(x)$ besides the solution $y(x)$ already found. Either solution can be substituted for any function $y_n(x)$ to satisfy the inequality (11.1), so we also obtain the inequality (11.4) for them, namely

$$|y(x) - \widehat{y}(x)| \leq (10^{-n} + 10^{-m}) \cdot \frac{e^{L(x-a)} - 1}{L}$$

If x_0 is any number in $[a, b]$, then because n and m can be taken arbitrarily large, the preceding inequality implies

$$|y(x_0) - \widehat{y}(x_0)| = 0$$

and the two solutions, supposedly distinct, are shown to be identical.

11.3 The Possibility of No Solution

In the previous section devoted to the initial value problem $y'(x) = f(x, y)$, $y(a) = y_0$, we showed that if $f(x, y)$ is uniformly continuous in a containment rectangle $R_{[a,b],M}$, then for any choice of ϵ, it is possible to construct a function $y(x)$ such that $|y'(x) - f(x, y(x))| < \epsilon$ for any x in $[a, b]$. In spite of this result, without the Lipschitz condition it is nevertheless possible that there is no *exact* solution to the initial value problem. This result contrasts with a standard result of conventional calculus known as Peano's theorem [32] [19, pp. 59–66], which asserts the existence of a solution if the function $f(x, y)$ is continuous in a containment rectangle.

THEOREM 11.2: Let R be the rectangle in the xy plane defined by the inequalities

$$0 \leq x \leq 1 \qquad -1 \leq y \leq 1$$

There exists a function $f(x, y)$ defined and uniformly continuous on R, such that there is no solution to the initial value problem $y' = f(x, y)$, $y(0) = 0$, on any interval $[0, h]$, no matter how close to 0 the positive constant h is.

PROOF: The proof consists of the construction of a function $f(x, y)$ with the following properties:

(a) $|f(x, y)| < 1$ for x in $[0, 1]$. This implies that over the x interval $[0, 1]$, the rectangle R is a containment rectangle for the given initial value problem, because as long as a solution $y(x)$ remains in R, we have $|y'(x)| = |f(x, y(x))| < 1$. But then $|y(x)| = |y(x) - y(0)| = |y'(c_x)x| < |1x|$, so for x in $[0, 1]$ the function $y(x)$ cannot leave R.

(b) $f(x, y)$ is uniformly continuous on R.

(c) If $y(x)$ is any solution to the differential equation $y'(x) = f(x, y)$ on some interval $[a, b]$ within $[0, 1]$, then the values $y(10^{-n})$ are all equal for positive integers n with 10^{-n} in $[a, b]$. Moreover, if $y(10^{-n})$ is zero, and both 10^{-n} and $3 \cdot 10^{-n}$ belong to $[a, b]$, then $y(3 \cdot 10^{-n})$ is $> \frac{1}{27}10^{-3n}$ if the nth program P defines a real number that is positive, and $y(3 \cdot 10^{-n})$ is $< -\frac{1}{27}10^{-3n}$ if the nth program P defines a real number that is negative. Here we are using the association of positive integers n with programs P first described in Section 4.2.

The condition (c) rules out any solution to the differential equation satisfying the initial condition $y(0) = 0$ in an interval $[0, h]$, no matter how small h is. If there were such a solution, then for arguments 10^{-n} in $[0, h]$ the value of $y(10^{-n})$ is unchanging, and this unchanging value must be 0, since $\lim_{n\to\infty} y(10^{-n}) = y(0) = 0$. If we had a solution $y(x)$ valid in some interval $[0, h]$, we could determine whether any real number a satisfied the condition $a \geq 0$ or satisfied the condition $0 \geq a$, contradicting Nonsolvable Problem 3.11. This would be the procedure: We would be able to find an integer N such that $10^{-N} < h$. We may assume the program P_a defining a has $<P_a>$ greater than N, because if this is not the case, we add to P_a sufficient additional steps that accomplish nothing until the revised P_a program satisfies this condition. Let D equal the integer $<P_a>$. With sufficient accuracy of our computation, we can determine for $c = y(3 \cdot 10^{-D})$ which one of the following 4 relations hold: $c > \frac{1}{27}10^{-3D}$, $c < \frac{1}{27}10^{-3D}$, $c > -\frac{1}{27}10^{-3D}$, $c < -\frac{1}{27}10^{-3D}$. In the first and third cases we know that $a \geq 0$. In the second and fourth cases, we know that $0 \geq a$. This contradicts Nonsolvable Problem 3.12, so our assumption of a solution in $[0, h]$ is wrong, and there can be no such solution.

To carry out the construction of $f(x, y)$, we first define a certain function $s(x, y)$ as follows:

$$s(x, y) = \begin{cases} 0 & \text{if } x < 1 \\ (x-1)(2-x)y^{1/3} & \text{if } x \text{ is in } [1, 2] \\ -(x-2)(3-x)y^{1/3} & \text{if } x \text{ is in } [2, 3] \\ 0 & \text{if } x > 3 \end{cases}$$

Consider the solution to the initial value problem

$$y'(x) = s(x, y), \qquad y(0) = y_0$$

The solution $y(x)$ will be constant and equal to y_0 for x in the interval $[0, 1]$, because $s(x, y)$ is zero for x in $[0, 1]$. In the x interval $[1, 2]$, the function $s(x, y)$ has a $y^{1/3}$ term which can cause multiple solutions, along with the factor $(x - 1)(2 - x)$, which also may be written as $\frac{1}{4} - (x - \frac{3}{2})^2$. This factor is 0 at $x = 1$, grows to $\frac{1}{4}$ at $x = \frac{3}{2}$, and then decreases symmetrically back to 0 at $x = 2$. Even with the factor present, the solutions obtained for various y_0 exhibit the same general behavior as found previously for the case $y' = y^{1/3}$, namely multiple solutions if the initial value y_0 is zero, and a single solution otherwise. Separating variables as before and solving, we obtain

$$\int_{y_0}^{y} \frac{dy}{y^{1/3}} = \int_{1}^{x} (x-1)(2-x)dx$$

$$\frac{3}{2}\left[y^{2/3} - y_0^{2/3}\right] = \frac{1}{2}(x-1)^2(2-x) + \frac{1}{6}(x-1)^3$$

$$y = \left[\pm\sqrt{y_0^{2/3} + \frac{1}{3}(x-1)^2(2-x) + \frac{1}{9}(x-1)^3}\,\right]^3$$

For the case $y_0 \neq 0$, the $\pm$ sign in the preceding equation is set to the sign of y_0. If y_0 is zero, again there is the zero solution, and two other solutions:

$$y(x) = \pm[\tfrac{1}{3}(x-1)^2(2-x) + \tfrac{1}{9}(x-1)^3]^{3/2}$$

Again, there are an infinite number of solutions obtainable by combining the zero solution with either of the other two solutions, properly shifted.

All these solutions can be distinquished by their value at $x = 2$. For solutions with y_0 nonzero, this is clear from the form of the solution. For solutions with y_0 zero, as soon as the solution leaves the line $y = 0$, it is a monotone function until x reaches 2, so the magnitude of $y(2)$ determines the point at which the solution left the line $y = 0$, and the sign of $y(2)$ indicates the direction the solution took when it left. We note in particular that the two solutions given in the preceding for the case $y_0 = 0$ have at $x = 2$ the values $\frac{1}{27}$ or $-\frac{1}{27}$ dependent upon whether the $\pm$ sign is $+$ or $-$.

Any solution $y(x)$ satisfies the identity

$$y(x) = y(4-x)$$

To verify this, note that the function $s(x, y)$ satisfies for all x the equation

$$s(x, y) = -s(4-x, y)$$

If $y(x)$ is a solution, and we set $y_1(x)$ equal to $y(4-x)$ and take its derivative, we obtain

$$y_1'(x) = \frac{d}{dx}y(4-x) = -y'(4-x) = -s_n(4-x, y(4-x))$$
$$= s(x, y(4-x)) = s(x, y_1(x))$$

Therefore, $y(4-x)$ is also a solution to the differential equation, and because $y(x)$ and $y(4-x)$ have the same value at $x = 2$, they are identical solutions.

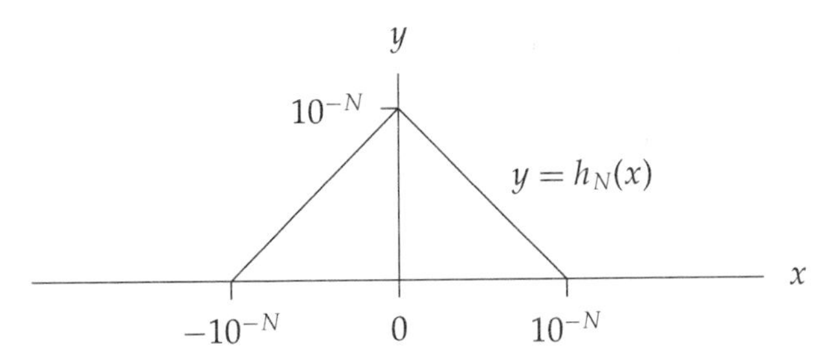

Figure 11.1. The spike function $h_N(x)$.

Now we use $s(x, y)$ to build a sequence of functions $s_n(x, y)$. The function $s_n(x, y)$ is assigned the nth program P, that is, the program P such that <P> equals n. The function $s_n(x, y)$ equals $s(x, y)$ except for certain changes made in the x intervals $[0, 1]$ and $[3, 4]$. The nth program P, presumed to be a D approximation function for a real number, is tested in an attempt to determine whether or not the number it defines equals zero. Taking the parameter D in succession equal to 1, 2, 3,..., we test the $P(D)$ interval values, attempting to find an interval that does not contain the zero point. If such an interval is found after following the computation chain through to exactly N steps, then $s_n(x, y)$ is altered by placing the spike function $h_N(x)$ (see Fig. 11.1) in the middle of its x interval $[0, 1]$, with the multiplier $+1$ if the deciding interval was positive, and with the multiplier -1 if the deciding interval was negative. The same spike with the opposite sign is placed in the middle of the x interval $[3, 4]$ of $s_n(x, y)$. That is, if it is determined that P defines a positive number, $s_n(x, y)$ is defined by the equation

$$s_n(x, y) = s(x, y) + h_N(x - \tfrac{1}{2}) - h_N(x - 3\tfrac{1}{2}) \tag{11.5}$$

If P defines a negative number, then $s_n(x, y)$ is defined by the equation

$$s_n(x, y) = s(x, y) - h_N(x - \tfrac{1}{2}) + h_N(x - 3\tfrac{1}{2}) \tag{11.6}$$

The purpose of putting opposite signed spikes in the intervals $[0, 1]$ and $[3, 4]$ is to preserve the relation $s(x, y) = -s(4 - x, y)$ for the function $s_n(x, y)$, in order that solutions to the differential equation $y'(x) = s_n(x, y)$ remain symmetric with respect to the line $x = 2$. This way a solution which is zero at $x = 0$ is again zero at $x = 4$.

To verify that $s_n(x, y)$ is computable, suppose x_0 is some number and we want a D approximation to $s_n(x_0, y)$. The s_n program need follow the P computation described only through D steps. If no conclusion has been reached by this number of steps, then if spikes are eventually added to $s(x, y)$, they are of height $10^{-(D+1)}$ or less, and it is safe to ignore such spikes and return a value for $s_n(x, y)$ by computing a $D+1$ value for $s(x, y)$ and adding on an extra error of $10^{-(D+1)}$. Of course, if a conclusion has been reached, then the $s(x, y)$ value is calculated by using Eq. (11.5) or Eq. (11.6).

We define the function $f(x, y)$ of the theorem by the equation

$$f(x, y) = \sum_{n=1}^{\infty} 10^{-2n} s_n(10^n(x - 10^{-n}), 10^{3n} y) \tag{11.7}$$

Under the transformation

$$\widehat{x} = 10^n(x - 10^{-n}), \qquad \widehat{y} = 10^{3n} y$$

with inverse

$$x = 10^{-n}\widehat{x} + 10^{-n}, \qquad y = 10^{-3n}\widehat{y}$$

the derivative $d\widehat{y}/d\widehat{x}$ becomes $10^{3n} \cdot 10^{-n}$ times the derivative dy/dx, so a solution to

$$\frac{d\widehat{y}}{d\widehat{x}} = s_n(\widehat{x}, \widehat{y})$$

for $\widehat{x}$ in $[0, 4]$, becomes a solution to

$$\frac{dy}{dx} = 10^{-2n} s_n(10^n(x - 10^{-n}), 10^{3n} y) \tag{11.8}$$

for x in the interval $[10^{-n}, 5 \cdot 10^{-n}]$. Moreover, for x outside the interval $[10^{-n}, 5 \cdot 10^{-n}]$, the right-hand side of Eq. (11.8) equals 0. Thus, for any argument (x, y), only one of the functions on the right-hand side of Eq. (11.7) can be unequal to zero. Also note that for (x, y) in the containment rectangle R of the theorem, the absolute value of the right-hand side of Eq. (11.8) does not exceed

$$10^{-2n} \cdot \tfrac{1}{4} \cdot (10^{3n})^{1/3} = \tfrac{1}{4} 10^{-n}$$

Hence $|f(x, y)| < 1$ as required.

The Corollary to Theorem 7.2, which concerns functions of one variable defined on intervals, can be generalized to apply to functions of two variables x and y defined on a rectangle R in the xy-plane. Using such a generalization of this corollary, we obtain that the sequence of functions

$$\sum_{n=1}^{N} 10^{-2n} s_n(10^n(x - 10^{-n}), 10^{3n} y) \tag{11.9}$$

converges uniformly on R to the limit function $f(x, y)$, and that $f(x, y)$ is uniformly continuous on R. This concludes the proof of the theorem.

■

CHAPTER

Ideal Computer Simulation

The ideal computer serves as the means of deciding whether any computation task can be done by finite means. Accordingly, we can expect to be able to program the ideal computer to accomplish much more than just what was demonstrated in Chapter 5. In this chapter we present subsidiary information about the ideal computer and the simulation software, along with two suggestions for possible programming projects.

12.1 The Extended Ideal Computer

We mentioned in Section 5.6 that in order to make it easier to simulate the ideal computer, four additional ideal computer step types are provided. These step types are used by the "extended ideal computer," which one may consider a peripheral device not central to the ideal computer concept. The extra step types are needed to prepare for a simulation run, to display the results of a simulation run, and to summarily stop ideal computer simulation.

We list the four additional step types:

Step type a: { Characters you enter at your keyboard become the contents of v_1, the <enter> key ending the input;
Step type b: The v_1 symbol string is displayed at your console;
Step type c: Carriage return signal is sent to your console; and
Step type q: Quit. (Immediate halt of simulation.)

The console interaction here is always with the ideal computer's action form and never the entry form.

When composing a source code program for the extended ideal computer, you can specify these additional steps only indirectly because the `Compile Source` software recognizes only the 11 basic ideal computer steps when it converts a source file into an execution file. The four extra steps are obtained through subroutine calls, specifically a call to `in_from_console` for step a, `out_to_console` for step b, `cr_to_console` for step c, and `abort` for step q. You can see the simple execution code each of these four programs has by loading the program into the ideal computer entry form.

12.2 Call Loops

For any ideal computer program P described in Chapter 5, one can draw a "directed line graph" diagram to represent the calling relationships between P and its subroutines. A vertex is assigned to P and to each subroutine used by P, and then if program P_1 calls program P_2, a directed line in the form of an arrow is drawn from the vertex representing P_1 to the vertex representing P_2, the arrow head being at the called program's vertex. The vertex assigned to the main program P is the "root" vertex. Such a diagram is shown in Fig. 12.1 for the program `add2integers`. Notice that in this diagram there is no directed line path that starts from the root vertex and returns again to this vertex. That implies that the program `add2integers` is "loop-free," that

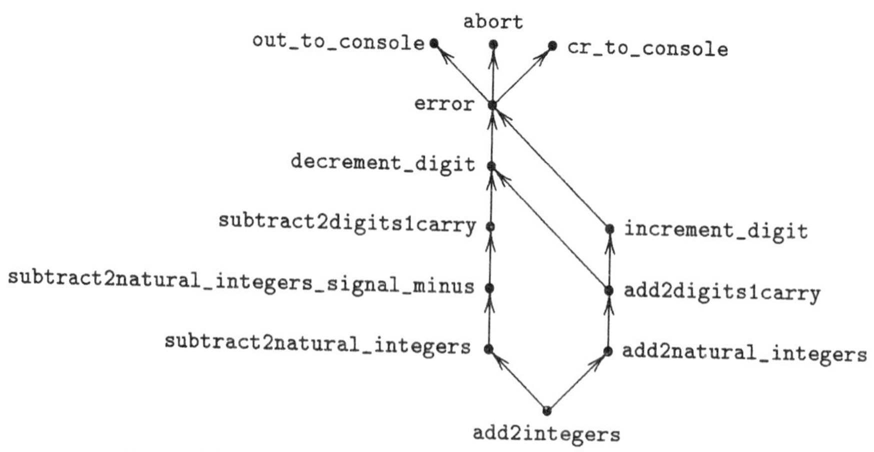

Figure 12.1. A graph of the subroutines used by `add2integers`.

is, never calls itself, either directly or through a sequence of subroutines. If a similar diagram is made for any of the programs so far described, a loop-free graph would be obtained.

In the definition of the ideal computer, there is no prohibition against a program calling itself, but there also is no essential advantage that is obtained thereby. We have generally avoided having call loops in order to have faster program execution. The subroutine call system requires using a list with an entry made for each successive call, the entry identifying the calling program, and its next step when it regains control. When program P_1 calls program P_2, an entry with this call information is appended to the list, and when control returns to P_1, the list entry is deleted. If there are call loops, then when program P_1 calls program P_2, the contents of P_1's variables must also be preserved on this list, because the program P_1 may eventually be called itself before P_2 returns control back to it. (Recall that when a program gets called, all noninput variables of the program are empty.) The `Execute` software checks the call graph of the target program to determine whether there are call loops, and if there are none, then a somewhat faster execution system is used whereby the variables of a calling program are *not* stored.

The ideal computer program `execute_program`, described in Section 5.22, simulates other ideal computer programs, and this program also uses a list to keep track of call chains. With `execute_program`, however, there is no difference in its handling of programs with call loops and programs without call loops.

The two programs `factorial_of_natural_integer` and `factorial_of_natural_integer_recursive_method` demonstrate contrasting ways of programming the function $n!$. The first program computes $n!$ in a straightforward fashion and has no call loops. The second program P' calls the subroutine `recursive_factorial`, with a v_1 input of n and a v_2 input of 1, and then `recursive_factorial` proceeds to generate $n!$ by repeatedly calling itself. If you load `recursive_factorial` into the entry form, you can see the method used by this subroutine. When `recursive_factorial` is called, it checks whether v_1 is nonzero, and if this is the case, multiplies the integer in v_2 by v_1, decrements v_1, calls itself to complete the computation, and returns. If v_1 is 0, `recursive_factorial` does no computation and just returns. Thus when `recursive_factorial` is first called by the second $n!$ program P' with, say, 10 in v_1 and 1 in v_2, a chain of 10 more calls of `recursive_factorial` ensues until finally v_1 becomes 0, and then each version of the subroutine returns control to the previous version, with control eventually returned to P'. The program P' then moves the $n!$ product in v_2 to v_1 and terminates.

12.3 Composing and Deleting a Program

It is easier to compose an ideal computer program that will be executable if you can adapt an existing executable program to your new purpose. This way you can use the `type` declaration and often some of the program steps as well. We illustrate the procedure by describing how to write a simple new program `example1` that makes use of the executable program `example`, first mentioned in Section 5.4.

Load the `example` program into the entry form and then make the `Options` menu selection `Copy Source`. A small form with `New Program` on its title bar appears, the form having the caption "`Enter ideal computer program name.`" Using your console keyboard, enter the name `example1`, then click on `OK`. The `example` source file is now copied into a newly created source file, `example1`.

You will be able now to load the `example1` program into the entry form. Notice that the `Options` menu choices `Execute` and `Execute Step by Step` are disabled. This is because the source file has not yet been compiled successfully. The first step in program modification is to change the first source line from `program_name = example` to `program_name = example1`. After you change a source file, if the file has not yet been compiled successfully, you can regain the original source file in the entry form by the `Options` menu selection `View Source`. For instance, if you make this selection after changing the `program_name` line, the old name `example` reappears. The source file is automatically saved after it is successfully compiled, at which time the corresponding type and execution code files get created if they are not already present, and then filled according to the source file's specification. You can also save the source file whenever you wish by the `File` menu selection `Save`.

We will make the `example1` program differ from `example` by duplicating the code line `v1 <- v1 + v2;`. The effect of this change will be to make the string returned by `example1` consist of the v_1 input string concatenated with a *duplicated* v_2 input string. This change to `example1` can be accomplished in two ways: You can click on the end of the code line to be duplicated, hit the `<Enter>` key, and then type in the code line. Or you can use the `Edit` menu selections `Copy` and `Paste`. Here you highlight the specified line by dragging the cursor over it, then you make the `Copy` selection, whereupon the highlighted text returns to its original state, signaling that the text has been copied. After this you can click on the end of the code line, hit the `<Enter>` key, and obtain the duplicated line with the menu selection `Paste`.

After you have made these changes to the `example1` source file, make the `Options` menu selection `Compile Source`. If the compilation is unsuccessful for some reason, you will obtain a displayed error report and the entry form will also show some character highlighted to indicate the point in the source file where the error occurred. (Perhaps the semicolon at the end of the code line was missing.) No error report indicates a successful compilation. Compilation often leads to changes in the `Options` menu display, and with our program `example1`, we find that now every `Options` menu selection is enabled, whereas previously half of them were disabled. You now can see `example1`'s execution code or type with the `View Code` or `View Type` menu selections. You also can now execute `example1` with the `Options` menu selections `Execute` or `Execute Step by Step`.

The procedure for removing `example1` from the ideal computer disk is also worth demonstrating. To eliminate all three of `example1`'s files, that is, the source, execution code, and type files, the first step is to delete completely the source file. If the entry form is not displaying the source file, regain the source file with the `Options` menu selection `View Source`. Next, in the entry form, click under the last displayed line, and then hold the backspace key down until the form is cleared. If you now make the `File` menu selection `Save`, the entry form's title bar should change from `example1` to `Ideal Computer Disk Files`, and this change indicates that all three `example1` files are removed. If the title bar does not change, this is because there still is a blank line or two in the `example1` source file. In this case, simply click a few lines down from the top of the apparently empty entry form, hold the backspace key down again, and then make the `File` menu selection `Save` a second time.

All ideal computer programs are stored on your hard disk in a directory with name `i_c_pgms`. This directory is located in the file folder that holds the ideal computer software (file name `IdealComputer.exe`). If you ever wish to return all program files to their original state, this is easily accomplished after you place the CD-ROM back into your CD drive. If you make the `File` menu selection `Delete all programs`, all program files are removed and the `File` menu choice `Load initial programs` becomes enabled. If you now make this second selection, the program files are recopied from the CD-ROM back to your hard disk.

The procedure just described can be modified in order to load the ideal computer with a set of starting programs differing from the ones on the CD-ROM. Perhaps you have composed a group of programs for some specialized use and want to be able to easily load the ideal computer with these programs. The CD-ROM holds its initial ideal computer programs in the CD-ROM directory `ideal`. When you make the menu selection `Load initial programs`, all programs residing in this directory are read into the `i_c_pgms` directory of your hard disk, whereupon they become accessible to the ideal computer. If your PC has a RW drive for CDs, then you can create a CD with your collection of programs in the CD's `ideal` directory. At any time afterwards, you can restart the ideal computer with your chosen set of programs. You need only insert the created CD in the CD drive, make the File menu selection Delete all programs and afterwards the selection Load initial programs.

12.4 Managing the Ideal Computer's Input and Output

For each executable program P of a particular type, a specific extended ideal computer program prepares P for execution by filling P's input variables appropriately, and after P terminates, another extended ideal computer program causes P's output to be displayed. For instance, programs of type rational_binary_operation utilize the program in2rationals for input preparation and the program out1variable for output display.

We identify the input preparation programs by always beginning their names with the symbols "in_," and identify the output handling programs by always beginning their names with the symbols "out_." An input preparation program often calls other programs as subroutines, and some of these programs, being used only for input preparation purposes, also have names with the prefix "in_". When you make the File menu selection Open, the displayed program names are listed alphabetically, so all input preparation programs are listed together in the "in_" section. Similarly, an output display program may call on other specialized output handling programs as subroutines, and when you make the Open selection, all output display programs get listed together in the "out_" section. Generally, the type of any "in_" or "out_" program is miscellaneous. We assign this type to any program not intended to be itself executable.

An input handling program often must monitor input that a user enters from a keyboard. Such input needs to be thoroughly checked for unintentional typos, otherwise the executed ideal computer program may enter its "undefined" realm. For example, the in_digit input handling program for increment_digit calls a subroutine that checks whether the digit obtained from a user's console is actually a decimal digit. The in_rational program makes a more elaborate check and, if there is an error, gives the position in the user's typed line where the error occurred.

We describe in general terms the steps to follow when creating a new, executable program type. It also may help in the discussion to give more details on how the Execute software performs its task. Suppose the new type is designated type_abc, the type's input program has the name in_abc, and the type's output handling program has the name out_abc. The console file has a list of type information for the use of the `Execute` software, with each executable type having a one-line entry in the file. To make `type_abc` current we must add the line shown next to the `console` file:

```
type_abc;in_abc;out_abc;
```

This line can be added anywhere in the `console` list. (The type lines in the `console` file are listed alphabetically by type name, but only for the purpose of visually locating type lines more easily when the file is loaded into the entry form.) The `Execute` software, when called on to simulate program `xyz` of this new type, does the following. It retrieves the type line from the `console` file, and uses this information to compose and store temporarily the following extended ideal computer program:

```
{execute>0<0;8in_abc;8xyz;8out_abc;q;}
```

Next the software retrieves all the segments of this new program (using a method similar to that of `name_to_program`), and then proceeds to follow the program steps (using a method much faster than that used by `execute_program`). In this way the software executes in sequence the three programs `in_abc`, `xyz`, and `out_abc`, with the `q` step ending the simulation.

Next consider the types assigned to semifunctions (Section 5.21) and to decision programs (Section 5.23), where a two-run execution system is used. Here the executed program creates another program, and only after the second program is executed is there output to display. If our new type, `type_abc`, is to be of this variety, we must choose a permanent name for the second program, say `created_abc`. Now the console entry needs three program names following the type heading:

```
type_abc;in_abc;created_abc;out_abc;
```

When the `Execute` software finds three files names when it looks up the `console` entry for a type, it makes two separate simulations. If the executed program of the new type has the name `xyz`, the first simulation run is for the program

```
{execute>0<0;8in_abc;8xyz;q;}
```

and the second simulation run is for the program

```
{execute>0<0;8created_abc;8out_abc;q;}
```

When composing a complex program that one does not intend to make executable, the usual type assigned, `miscellaneous`, prevents any test of the program via the `Execute Step by Step` menu selection. However, stepwise execution of a program is often the quickest way of locating a program bug, so it may be helpful here to describe how to temporarily obtain the ability to step through any newly composed program. When a program is loaded into the entry form, at this point the program's type is checked. If the type is `miscellaneous` or `program_template`, and also if a type file is not present, the `Options` menu selections `Execute` and `Execute Step by Step` are disabled. The first step in making a `miscellaneous` program executable is to temporarily change the program's type. Any new unused type name will do; suppose the type line in the source file is changed to `miscellaneous1`, perhaps to make it easier to restore the type later to `miscellaneous`. The program will have its type file showing the new type after we make the `Options` menu selection `Compile Source`.

The next step is to load the `console` file into the entry form and add to the end of the file the line:

```
miscellaneous1;in_2strings;out_1variable;
```

The `in_2strings` program makes no examination of a user's typed input. Here we have assumed that the newly created program receives two input arguments and returns one argument, so the displayed type line is similar to the type line for `string_binary_operation`. If the program under test receives a different number of input arguments, let us say 3, we will need to write an appropriate string input program by copying `in_2strings` into `in_3strings` (as described in Section 12.3), and then modifying the copied program appropriately. The `console` type line then must have `in_2strings` replaced by `in_3strings`. Similarly, if the program under test returns more than one argument, the `out_1variable` entry in the type line would be changed accordingly.

After these changes are made, the program under test can be loaded into the entry form and executed step by step. When testing is complete, the program's type can be reset to `miscellaneous` and the program then recompiled.

Note that the foregoing directions do not apply to `miscellaneous` type programs that call for console input, because of the necessity here of retaining the usual keyboard action, rather than having every key depression initiate the advance to the next step, as occurs with `Execute Step by Step`.

12.5 Possible Projects

We describe here two programming projects in order to encourage a reader to modify the ideal computer programs for other purposes and also to compose new programs.

i. Conversion from the Decimal System to Another System

We have used the decimal system for our representation of integers, whereas most modern commercial computers calculate in the binary system. Sometimes these computer generated integers are represented in a 2-power system different from the binary system, such as the octal system or the hexadecimal system.

There are only three key programs that need to be changed in order to switch the ideal computer from the decimal system to another system. However, the switch from one system to another system should be done only after the ideal computer programs are restored to their initial state, as explained in Section 12.3. The restoration deletes all the computed numbers in the old system.

After the ideal computer programs are brought to their initial state, the three key programs can be changed, the effect caused by these changes explored, and then the original programs restored once more. The three programs that need to be changed are `increment_digit`, `decrement_digit`, and `check_digit`. The `check_digit` program, not previously mentioned, is used by input handling programs to verify that every keyed-in digit is a decimal digit.

To convert to the binary system, make changes to these programs by deleting much of the source code and making the obvious adjustments. The digit action code lines for `increment_digit` and `decrement_digit` will be identical, the binary system being the only system for which this occurs. After these three programs are changed and compiled, when you now execute a program such as `add2integers` or `divide2rationals`, you must type in binary quantities or else receive an error report.

There are a small number of programs that use the digit 2 in some capacity in their procedure, such as the programs `two_d`, `sqrt2_d`, `real_type_e_to_real_type_d`, and `nth_root_of_real_type_d`. When executed, these programs will report an input error with `decrement_digit` or `increment_digit`. These programs execute correctly in any system having the digit 2. Thus, if you convert to the ternary system, the octal system, or the hexadecimal system, by making appropriate changes to the three key programs mentioned, you will be able to make more extensive computations than you can in the binary system. (The hexadecimal system requires *adding* code lines to each of the three key programs in order to obtain proper action for the six extra hexadecimal digits a, b, c, d, e, and f, needed to represent, respectively, digits having a decimal value of 10, 11, 12, 13, 14, and 15.)

The `pi_e` program does not execute correctly in any system besides the decimal system. This is because the program makes use of several decimally specified rational numbers. However, pi_e will execute correctly in any system having the digit 2, if these rationals are expressed in that system.

ii. Writing a New Program

Suppose we want a program to compute binomial coefficients. We suggest this objective mainly to provide a concrete goal, and also because there is a program available to compute $n!$. The binomial coefficient formula is

$$\binom{n}{i} = \frac{n!}{i!\,(n-i)!}$$

If $n < i$, we take $\binom{n}{i}$ to be 0.

We need to choose a name for the program, and also choose a type. When we search through the `console` file, we find a useable type, namely `natural_integer_binary_operation`. If we use this type, then when we execute the program, we automatically obtain two natural integers in v_1 and v_2. Let us take the v_1 input value to be n and the v_2 input value to be i. We see that our program will need to call as subroutines `subtract2natural_integers_signal_minus` and `factorial_of_natural_integer`.

To get started with the program, we make the `File` menu selection `New`. A small form appears requesting a name for the new program. After we enter a name, the entry form shows the beginning source file for the program, with the `program_name` line already filled in. After we write the necessary code lines to generate the binomial coefficient, we are ready to make the `Options` menu selection `Compile Source`. Hopefully there will be no error report, but if there is one, the source of the error is highlighted in the entry form. No error report means the compilation was successful and the program is now executable. In that case we can now execute the program step by step and check the various steps we wrote.

References

[1] Aberth, O. (1968). Analysis in the computable number field, *J. Assoc. Comput. Mach.* **15:**275–299.

[2] Aberth, O. (1969). A chain of inclusion relations in computable analysis, *Proc. Amer. Math. Soc.* **22:**539–548.

[3] Aberth, O. (1971). The failure in computable analysis of a classical existence theorem for differential equations, *Proc. Amer. Math. Soc.* **30:**151–156.

[4] Aberth, O. (1980). *Computable Analysis.* New York: McGraw-Hill.

[5] Aberth, O. (1998). *Precise Numerical Methods Using C++.* Boston: Academic Press.

[6] Alefeld, G. and Herzberger, J. (1983). *Introduction to Interval Computation,* Translated by Jon Rokne. New York: Academic Press.

[7] Ceitin, G. S. (1955). On the theorem of Cauchy in constructive analysis, *Uspehi Mat. Nauk* **10:**207–209.

[8] Ceitin, G. S. (1959). Algorithmic operators in constructive complete separable metric spaces, *Dokl. Akad. Nauk SSSR* **128:**49–52.

[9] Ceitin, G. S. (1962). Algorithmic operators in constructive metric spaces, *Trudy Mat. Inst. Steklov* **67:**295–361; (1967). English trans., *Amer. Math. Soc. Trans., series 2* **64:**1–80.

[10] Ceitin, G. S. (1962). Mean value theorems in constructive analysis, *Trudy Mat. Inst. Steklov* **67:**362–384; (1971). English trans., *Amer. Math. Soc. Trans., series 2* **98:**11–40.

[11] Ceitin, G. S. (1964). Three theorems in constructive functions, *Trudy Mat. Inst. Steklov* **72:**537–543; (1972). English trans., *Amer. Math. Soc. Trans., series 2* **100:**201–209.

[12] Church, A. (1936). A note on the Entscheidungsproblem, *The Journal of Symbolic Logic* **1:**40–41, 101–102.

[13] Davis, M. (1958). *Computability and Unsolvability*. New York: McGraw-Hill.

[14] Demut, O. (1967). Necessary and sufficient conditions for the Riemann integrability of constructive functions, *Dokl. Akad. Nauk SSSR* **176:**757–758; (1967). English trans., *Soviet Math. Dokl.* **8:**1176–1177.

[15] Epstein, R. L. and Carnielli, W. A. (1989). *Computability: Computable Functions, Logic, and the Foundations of Mathematics*. Pacific Grove, California: Wadworth & Brooks/Cole.

[16] Goodstein, R. L. (1961). *Recursive Analysis*. Amsterdam: North-Holland.

[17] Gödel, K. (1931). Über formal unentscheidbare Sätze der Principia Mathematica und vervandter Systeme I, *Monatshefte für Mathematik und Physik* **38:**173–198.

[18] Jolley, L. B. W. (1961). *Summation of Series*, 2nd revised ed., New York: Dover.

[19] Kamke, E. (1943). *Differentialgleichungen reeler Funkionen*. New York: Chelsea.

[20] Kushner, B. A. (1964). Riemann integration in constructive analysis, *Dokl. Akad. Nauk SSSR* **156:**255–257; (1964). English trans., *Soviet Math. Dokl.* **5:**628–630.

[21] Kushner, B. A. (1965). On the existence of unbounded analytic constructive functions, *Dokl. Akad. Nauk SSSR* **160:**29–31; (1965). English trans., *Soviet Math. Dokl.* **6:**26–28.

[22] Kushner, B. A. (1965). On the constructive theory of Riemann integration, *Dokl. Akad. Nauk SSSR* **165:**1238–1240; (1965). English trans., *Soviet Math. Dokl.* **6:**1584–1587.

[23] Kushner, B. A. (1980). *Lectures on Constructive Mathematical Analysis*, Translations of Mathematical Monographs, Vol. 60. Providence, R. I.: American Mathematical Society.

[24] Markov. A. A. (1954). On the continuity of constructive functions, *Uspehi Mat. Nauk* **9:**226–230.

[25] Markov. A. A. (1958). On constructive functions, *Trudy Mat. Inst. Steklov* **52:**315–348.

[26] Markov. A. A. (1962). On constructive mathematics, *Trudy Mat. Inst. Steklov* **67:**8–14.

[27] Moore, R. E. (1962). *Interval Arithmetic and Automatic Error Analysis in Digital Computing*, Ph. D. Dissertation, Stanford University.

[28] Moore, R. E. (1966). *Interval Analysis*. Englewood Cliffs, NJ: Prentice-Hall.

[29] Moore, R. E. (1979). *Methods and Applications of Interval Analysis, SIAM Studies in Applied Mathematics*. Philadelphia: SIAM.

[30] Neumaier, A. (1990). *Interval Methods of Systems of Equations, Encyclopedia of Mathematics and Its Applications*, Cambridge: Cambridge University Press.

[31] Orevkov, V. P. (1963). A constructive map of the square into itself which moves every constructive point, *Dokl. Akad. Nauk SSSR* **152:**55–58; (1963). English trans., *Soviet Math. Dokl.* **4:**1253–1256.

[32] Peano, G. (1890). Démonstration de l'intégrabilité des équations différentielles ordinaires, *Math. Ann.* **37:**182–228.

[33] Rice, H. G. (1954). Recursive real numbers, *Proc. Amer. Math. Soc.* **5:**784–791.

[34] Specker, E. (1949). Nicht Konstruktiv Beweisbare Satze der Analysis, *Journal of Symbolic Logic* **14:**145–158.

[35] Specker, E. (1959). Der Satz vom Maximum in der Rekursive Analysis, in *Constructivity in Mathematics*, Proc. Colloq. at Amsterdam 1957, 254–265. Amsterdam: North Holland.

[36] Turing, A. M. (1937). On computable numbers, with an application to the Entscheidungsproblem, *Proc. Lond. Math. Soc., series 2* **42:**230–265.

[37] Turing, A. M. (1937). A correction, *Proc. Lond. Math. Soc., series 2* **43:**544–546.

[38] Widder, D. V. (1947). *Advanced Calculus*. New York: Prentice-Hall.

[39] Zaslavsky, I. D. (1955). The refutation of some theorems of classical analysis in constructive analysis, *Uspehi Mat. Nauk* **10:**209–210.

[40] Zaslavsky, I. D. and Ceitin, G. S. (1956). On the relations among the fundamental properties of constructive functions, *Proc. Third All-Union Math. Congr. Acad. Sci. USSR 1956*, 180–181.

[41] Zaslavsky, I. D. (1956). Some peculiarities of constructive functions of a real variable in comparison with classical functions, *Proc. Third All-Union Math. Congr. Acad. Sci. USSR 1956*, 181–182.

[42] Zaslavsky, I. D. (1962). Some properties of constructive real numbers and constructive functions, *Trudy Math. Inst. Steklov* **67**:385–457; (1966). English trans., *Amer. Soc. Trans., series 2* **57**:1–84.

[43] Zaslavsky, I. D. and Ceitin, G. S. (1962). Singular coverings and properties of constructive functions connected with them, *Trudy Math. Inst. Steklov* **67**:458–502; (1971). English trans., *Amer. Math. Soc. Trans., series 2* **98**:41–89.

[44] Zaslavsky, I. D. (1964). Differentiation and integration of constructive functions, *Dokl. Akad. Nauk SSSR* **156**:25–27; (1964). English trans., *Soviet Math. Dokl.* **5**:599–601.

About the CD-ROM Accompanying *Computable Calculus*

The software on the CD-ROM for simulating the Ideal Computer described in the text is for a PC using Windows 95 or a later Microsoft operating system.

Loading the Software

Load the CD into your PC's CD drive. Then from the Windows `Start` menu, choose the `Run` option, and if your CD drive is assigned the letter '`d`', enter the command `d:setup`. (Change the prefix letter '`d`' appropriately if your CD drive has some other letter assigned.) After the setup process is complete, click on the Ideal Computer in your Windows `Programs` display. The menu on the form which appears will have, in the `File` section, the command: `Load initial programs`. Execute this command to load all the initial Ideal Computer programs contained on the CD. These programs are all small in length and take up together less than half of a megabyte of your hard disk space. Chapter 5 has a detailed description of the Ideal Computer, and the explanation of how to use the computer begins in Section 5.5.

Removing the Software

Should you ever wish to remove the Ideal Computer software, you need to reverse the steps described above. First remove all the loaded Ideal Computer programs by clicking on the Ideal Computer in your Windows `Programs` display, and then execute the `File` menu command: `Delete all programs`. After this is done, return to your Windows `Start` menu, choose `Settings`, and then choose `Control Panel`. Among your choices now will be: `Add/Remove Programs`, and taking this option will allow you to designate the Ideal Computer for removal. Afterwards, your Windows `Programs` display may still show the Ideal Computer, but not after you shut down your computer and start it up again.

Index